Intellectual Property Righ
Objective Analysis

The Authors

Phundan Singh [b. 1946] hails from a reputed agricultural family of Western Uttar Pradesh [Village-Sakauti, Tehsil- Mawana, District- Meerut]. He graduated in Agriculture from Agra University, Agra and obtained his Master and Doctorate degrees from Kanpur University. He has throughout a brilliant academic record. His area of specialization is Plant Breeding and Genetics. He was selected as Scientist of first competitive examination conducted by ICAR through ASRB.

Dr. P. Singh has total experience of more than 40 years in the field of Plant Breeding and Genetics. He has served the Central Institute for Cotton Research, Nagpur for about 32 years in various capacities such as Scientist, Senior Scientist, Principal Scientist, Head, Division of Crop Improvement and Director. He has about 300 publications to his credit. He has participated and presented 80 research papers in various National and International Seminars, Symposia and Conferences. He has authored 52 books mostly published by Kalyani Publishers, New Delhi, 18 Scientific bulletins and contributed 14 chapters in various reputed books.

Dr. P. Singh is a member of several Scientific Societies, referee of different journals, Expert member of Selection Committees, M. Sc. and Ph.D. Examiner and paper setter in different Agricultural Universities. The CICR was recognized as the referral Laboratory for detection of Bt. gene and received the best Annual report Award in 2003 when he was the Director. He is a widely travelled Scientist. He has visited several countries such as Russia, Belarus, Ukraine, Crimea, Uzbekistan, Canada, USA and England.

Engineer Rajeev Singh graduated in Mechanical Engineering from Visvesvaraya National Institute of Technology, Nagpur in 1st Division. Based on campus interview, he was selected as Assistant Manager by Reliance Industries Limited (RIL). Later on he was elevated to the post of Manager, Senior Manager, General Manager and Senior General Manager. Presently, he is working as Senior General Manager with Reliance Industries Limited. He has throughout a brilliant academic record. He has more than 27 year managerial experience. He has authored one book entitled IPR and Plant Breeders' Rights.

He has visited two countries, *viz.* United States of America and Republic of China for official work. He obtained Gold Medal in shooting competition at Surat organized by Surat City Rifle Club. He obtained Certificate in Rifle Shooting from Swarnim Gujarat Sports University and got Gold Medal topping the list of successful candidates. His hobbies are running, cycling, shooting, *etc.*

Intellectual Property Rights
Objective Analysis

— Authors —

Phundan Singh

Former Director
Central Institute for Cotton Research,
Nagpur – 440 010
Maharashtra, India
Mahatma Phule Krishi Vidyapeeth, Rahuri, Maharashtra, India

and

Er. Rajeev Singh

Senior General Manager
Reliance Industries Limited,
Surat (Gujarat)

2024

Daya Publishing House®

A Division of

Astral International Pvt. Ltd.

New Delhi – 110 002

ISBN: 978-93-5124-886-6 (HB)

Reprinted 2024

Publisher's Note:

Every possible effort has been made to ensure that the information contained in this book is accurate at the time of going to press, and the publisher and author cannot accept responsibility for any errors or omissions, however caused. No responsibility for loss or damage occasioned to any person acting, or refraining from action, as a result of the material in this publication can be accepted by the editor, the publisher or the author. The Publisher is not associated with any product or vendor mentioned in the book. The contents of this work are intended to further general scientific research, understanding and discussion only. Readers should consult with a specialist where appropriate.

Every effort has been made to trace the owners of copyright material used in this book, if any. The author and the publisher will be grateful for any omission brought to their notice for acknowledgement in the future editions of the book.

Published by : **Daya Publishing House®**
A Division of
Astral International Pvt. Ltd.
– ISO 9001:2015 Certified Company –
4736/23, Ansari Road, Darya Ganj
New Delhi-110 002
Ph. 011-43549197, 23278134
E-mail: info@astralint.com
Website: www.astralint.com

Digitally Printed at : Replika Press Pvt. Ltd.

Dedicated to Grandchildren of
Senior Author

Miss Shubhika Singh
and
Master Rishabh Singh

Preface

The subject of Intellectual Property Rights is gaining increasing importance these days. There are several books on this subject both of foreign and Indian origin. However, almost all these books deal with the subject in a descriptive manner. There is hardly any book which treats the subject in objective manner. The prime objective of writing this book is to fill up this gap and present the subject in objective manner. This book has been designed for competitive examinations such as JRF, SRF, NET, SET, bank officers, agriculture officers, *etc.*

This book contains 21 chapters which have been divided into two sections. Section I deals with Primary Rights comprising 12 chapters [Chapter 1 to Chapter 12] and Section II deals with Sui-generic Rights consisting of 9 Chapters [Chapter 13 to Chapter 21]. The information contained in this book has been gathered from various published sources and Internet websites. Attempts have been made to provide latest information,even then some valuable information might have been missed.

The summary of Intellectual Property Rights, addresses of Copyrights, Patent and trademarks Offices in India and lists of domain names of some 50 countries have been appended. The Glossary of important technical terms used in relation to Intellectual Property Rights has been provided at the end for ready reference. Important References have been listed at the end which would be useful in gathering further details.

It requires long hours to complete job like this. Completion of this manuscript took more than three years. I could not spare adequate time for my family members during writing of this book for which I felt guilty. The cooperation extended by my wife Mrs. Jaswanti Singh during preparation of this Volume is highly appreciable

Hope this Volume would be useful to the students, researchers, Plan Breeders, Design developers, artists, musicians, private seed companies, *etc.* Constructive suggestions of readers are welcome for further improvement of this book.

Phundan Singh

Rajeev Singh

Contents

Section I
Primary Rights

Chapter 1

Outlines of Intellectual Property Rights

A. Definitions

Copyright: The exclusive rights granted to author/creator to reproduce, sell and distribute a work, prepare derivative works and display the work publicly.

Domain Name: The names and words that companies designate for their registered Internet Web site addresses.

Exclusive Rights: The legal privileges that only a copyright owner has with respect to their copyrighted work.

Geographical Indication: A geographical indication is a sign used on goods that have a specific geographical origin and often possess qualities or a reputation that are due to that place of origin.

Immovable Property: Fixed types of properties which cannot move from one place to other. Such properties include land, buildings [house, flat, bungalow, farm house] and gardens.

Industrial Design: Any original shape, picture, or some combination applied to a useful article of manufacture. In other words, it refers tothe design of the mass-produced products of our everyday environment, from sinks and furniture to computers.

Intellectual Property: The product/process/idea which is outcome of the brain of a person and can be used on commercial scale for benefit of human kind.

Intellectual Property Rights: The legal rights provided to an inventor to derive economic benefits from his invention/innovation.

Movable Property: The property which can be easily shifted from one place to other. Such property includes animals, farm machines, furniture, fixtures, gold, silver, diamond, and money.

Patent: A document granting an inventor sole rights to an invention. It is an official document which grants sole rights to the inventor for manufacturing and marketing his product/process/invention to derive benefits.

Property: The wealth or valuable things earned by a person.

Trademark: A trademark can be a word, name, symbol, device or mark which is used to identify and distinguish the goods or services of one company from goods or services of another.

Trade Name: The name under which a company conducts business, or by which its business, product/service are identified. It may or may not be registered as a trademark.

Trade Secret: A formula or process or device used in business that is not published or divulged which gives an advantage over competitors.

B. Background Information

1. The property refers to wealth or valuable things earned by a person. It is estimated in terms of land, house, garden, industries, animals, gold, silver, diamond, money *etc.* accumulated by a person.

2. Each person wants to earn property to get a better or higher status in the society, to lead a better quality of life in the society, to get a name and fame in the society. In the society, rich people get more respect than others and for better financial security of life for himself and for his children.

3. Property is of three types, *viz.*, (i) immovable property, (ii) movable property and intellectual property.

4. Immovable Property include land, buildings [house, flat, bungalow, farm house] and gardens. Such properties are protected by laws of land in each country.

5. Movable Property includes animals, farm machines, furniture, fixtures, gold, silver, diamond, and money. Such properties are also protected by public laws.

6. Intellectual Property includes product/process/idea which is outcome of the brain of a person and can be used on commercial scale for benefit of human kind. It includes: inventions, literary and artistic works, and symbols, names, images, and designs used on commercial scale.

7. IP is measured in terms of new ideas, processes, products, inventions and innovations developed by a person.

8. IP requires lot of intellectual inputs in terms of thinking, planning and fine tuning of new ideas/products/processes, *etc.*

9. IP requires considerable amount of funds and other resources to develop new products/processes.

10. The main problem with intellectual property is that it can be copied, reproduced and used by others resulting in loss of inventor.

11. Intellectual property may be protected by copyright, trademark or patent. The holder of intellectual property rights is usually the person or persons who developed the product or the organization that funded it.

12. The Intellectual Property Rights are broadly divided into two groups, *viz.*, Primary Rights and *Sui-generis* Rights.

13. Primary IP rights include,Copyrights, Patents, Trademarks, Trade Secret, Trade names, Domain names, Industrial Designs,Geographical Indications, *etc.*

14. *Sui-generis* refers to things of their own kind or things with unique characteristics. Such rights include Database Rights, Mask work, Plant Breeders Rights, Traditional Knowledge, Moral Rights and Supplementary Protection Certificate.

15. Copyrights refer to the legal right granted to an author, composer, playwright, publisher, or distributor to exclusive publication, production, sale, or distribution of a literary, musical, dramatic, or artistic work is called copyright.

16. Patent is an official document which grants sole rights to the inventor for manufacturing and marketing his product/process/invention to derive benefits.

17. A trademark or trademark is a distinctive sign of some kind which is used by an individual, business organization or other legal entity to uniquely identify the source of its products and/or services to consumers, and to distinguish its products or services from those of other entities.

18. An industrial design is the ornamental or aesthetic aspect of an article produced by industry or handicraft.

19. A utility model is an exclusive right granted for an invention, which allows the right holder to prevent others from commercially using the protected invention, without his authorization, for a limited period of time.

20. A geographical indication is a sign used on goods that have a specific geographical origin and often possess qualities or a reputation that are due to that place of origin.

21. A trade secret is anything [a formula, process, method, mechanism, tool, pattern or device] which the disclosing party desires to keep secret.

22. Related rights are rights which are similar to authors' right but which are not connected with the actual author of the work. Both authors' rights and related rights are copyrights.

23. A trade name is the name under which a product is commercially known. The commercial name by which a chemical is known is called trade name. One chemical may have a variety of trade names depending on the manufacturers or distributors involved.

24. Domain refers to the unique name that identifies an Internet site. Domain names have two or more parts separated by dots. For example www.ask-edi.com or **www.bcbsks.com**

25. IP rights promote healthy competition for invention/innovation among the intellectuals and helps in improving the quality of the product. It makes available new ideas/technologies to different countries and it leads to faster development of industries/organizations engaged in research and development work.

26. Intellectual Property Rights involve lot of money transaction in registration, renewal and licensing. It invites lots of court cases due to infringements and may lead to monopoly of right holders, *etc.*

C. Multiple Choice Questions

1. **Which of the following is movable property?**

(a) Land (b) House
(c) Gold (d) Garden

2. **Which of the following is movable property?**

(a) Bungalow (b) House
(c) Industries (d) Diamond

3. **Which of the following is movable property?**

(a) Land (b) House
(c) Animals (d) Garden

4. **Which of the following is immovable property?**

(a) Silver (b) House
(c) Gold (d) Diamond

5. **Which of the following is immovable property?**
   (a) Land (b) Silver
   (c) Gold (d) Diamond

6. **Which of the following is immovable property?**
   (a) Garden (b) Silver
   (c) Gold (d) Diamond

7. **Which of the following is immovable property?**
   (a) Product (b) House
   (c) Idea (d) Diamond

8. **Which of the following is immovable property?**
   (a) Silver (b) Bungalow
   (c) Gold (d) Diamond

9. **Which of the following is an intellectual property?**
   (a) Silver (b) Patent
   (c) Gold (d) Diamond

10. **Which of the following is an intellectual property?**
    (a) Copyright (b) House
    (c) Gold (d) Diamond

11. **Which of the following is an intellectual property?**
    (a) Trademark (b) House
    (c) Gold (d) Diamond

12. **Which of the following is an intellectual property?**
    (a) Trade secret (b) House
    (c) Gold (d) Diamond

13. **Which of the following is an intellectual property?**
    (a) Invention (b) House
    (c) Gold (d) Diamond

14. **Which of the following is an intellectual property?**
    (a) Trade Name (b) House
    (c) Gold (d) Diamond

15. **Which of the following is not an intellectual property?**
    (a) Copyright (b) Patent
    (c) Domain name (d) Diamond

16. Which of the following is not an intellectual property?

(a) Copyright (b) Patent

(c) Trade Secret (d) Diamond

17. Which of the following is not an intellectual property?

(a) Copyright (b) Patent

(c) Gold (d) Trademark

18. Which of the following is not an intellectual property?

(a) Copyright (b) Patent

(c) Moral right (d) Diamond

19. Which of the following is not a primary right?

(a) Copyright (b) Patent

(c) Trade Secret (d) Moral right

20. Which of the following is not a primary right?

(a) Copyright (b) Patent

(c) Database right (d) Trademark

D. True or False

1. Land is a form of immovable property.
2. Garden is form of immovable property.
3. Bungalow is form of immovable property.
4. Gold is a form of movable property.
5. Silver is a form of movable property.
6. Diamond is a form of movable property.
7. Currency is a form of movable property.
8. Farm house is form of immovable property.
9. Animals are forms of movable property.
10. Car is a form of movable property.
11. Wealth earned by a person is called property.
12. Creations of mind refer to intellectual property.
13. Invention is a form of intellectual property.
14. Copyright is a form of intellectual property.

15. Patent is a form of intellectual property.
16. Trademark is a form of intellectual property.
17. A new process is a form of intellectual property.
18. A new idea of commercial use refers to intellectual pro property.
19. Plant breeders' right come under *sui-generis* rights.
20. Farmers' rights come under *sui-generis* rights.

E. Fill in the Blanks

1. Land is a form of ________ property.
2. Garden is form of ________ property.
3. Bungalow is form of ________ property.
4. Gold is a form of ________ property.
5. Silver is a form of ________ property.
6. Diamond is a form of ________ property.
7. Currency is a form of ________ property.
8. Farm house is form of ________ property.
9. Animals are forms of ________ property.
10. Car is a form of ________ property.
11. Wealth earned by a person is called ________.
12. Creations of mind refer to ________ property.
13. Invention is a form of ________ property.
14. Copyright is a form of ________ property.
15. Patent is a form of ________ property.
16. Trade mark is a form of ________ property.
17. A new process is a form of ________ property.
18. A new idea of commercial use refers to ________ property.
19. Plant breeders' right come under ________ rights.
20. Farmers' rights come under ________ rights.

F. Answer in Short or One Word

1. The wealth or valuable things earned by a person.
2. Fixed types of properties which cannot be moved from one place to other.
3. The property which can be easily shifted from one place to other.
4. The product/process/idea which is outcome of the brain of a person and can be used on commercial scale for benefit of human kind.
5. The legal rights provided to an inventor to derive economic benefits from his invention/innovation.
6. The exclusive rights to reproduce, sell and distribute a work, prepare derivative works and display the work publicly.
7. An official document which grants sole rights to the inventor for manufacturing and marketing his product/process/invention to derive benefits.
8. A word, name, symbol, device or mark which is used to identify and distinguish the goods or services of one company from goods or services of another.
9. A formula or process or device used in business that is not published or divulged which gives an advantage over competitors.
10. The name under which a company conducts business, or by which its business, product/service are identified.
11. The names and words that companies designate for their registered Internet Web site addresses.
12. Any original shape, picture, or some combination applied to a useful article of manufacture.
13. A sign used on goods that have a specific geographical origin and often possess qualities or a reputation that are due to that place of origin.
14. The legal privileges that only a copyright owner has with respect to their copyrighted work.

ANSWERS

C. Multiple Choice Questions

Sl.No.	*Answer*	*Sl.No.*	*Answer*
1.	(c) Gold	**11.**	(a) Trademark
2.	(d) Diamond	**12.**	(a) Trade secret
3.	(c) Animals	**13.**	(a) Invention
4.	(b) House	**14.**	(a) Trade Name
5.	(a) Land	**15.**	(d) Diamond
6.	(a) Garden	**16.**	(d) Diamond
7.	(b) House	**17.**	(c) Gold
8.	(b) Bungalow	**18.**	(d) Diamond
9.	(b) Patent	**19.**	(d) Moral right
10.	(a) Copyright	**20.**	(c) Database right

D. True or False

1 to 20 all are true

E. Fill in the Blanks

Sl.No.	*Answer*	*Sl.No.*	*Answer*
1.	Immovable	**11.**	Property
2.	Immovable	**12.**	Intellectual
3.	Immovable	**13.**	Intellectual
4.	Movable	**14.**	Intellectual
5.	Movable	**15.**	Intellectual
6.	Movable	**16.**	Intellectual
7.	Movable	**17.**	Intellectual
8.	Immovable	**18.**	Intellectual
9.	Movable	**19.**	*Sui-generis*
10.	Movable	**20.**	*Sui-generis*

F. Short or One Word Answer

Sl.No.	*Answer*	*Sl.No.*	*Answer*
1.	Property	**8.**	Trademark
2.	Immovable property	**9.**	Trade Secret

3. Movable property
4. Intellectual property
5. Intellectual property rights
6. Copyright
7. Patent
10. Trade Name
11. Domain Name
12. Industrial Design
13. Geographical indication
14. Exclusive rights

CHAPTER 2

Copyrights

A. Definitions

Copyright: The exclusive rights granted to author/creator to reproduce, sell and distribute a work, prepare derivative works and display the work publicly.

Infringement: An unauthorized use of material protected by copyright, patent or trademark law.

Fair us: The privilege to use the copyrighted material in a reasonable manner without the consent of the owner.

Hire Work: A work which is created by an employee in the regular course of his employment,

Plagiarism: Copying another's words.

B. Background Information

1. Copyright refers to the legal right granted to an author, composer, playwright, publisher, or distributor to exclusive publication, production, sale, or distribution of a literary, musical, dramatic, or artistic work.

2. **Period:** It provides protection for a specific duration. In India, the duration of copyright is 60 years after the death of the Author. For a work for hire or for an anonymous or pseudonymous work, the copyright lasts for 95 years after publication or 120 years after creation whichever is earlier.

3. **Registration:** The registration is not compulsory for copyright. In other words, it is an unregistered right. Since there is no registration for copyright, there is no need of filling application and there is no fee for copyright. However, it is advisable to register the copyright with the copyright office. It will provide additional legal rights to the copyright holder.

4. **Validity:** The copyright is valid in all the Countries of the World.
5. **Transfer:** The copyright holder has the rights to authorize others to use the protected work. In other words, the author may transfer the copyright to any other party if he so desires.
6. **Items Covered:** It is applicable to books, movies, music, paintings, photographs, sound recordings and software.
7. **Control:** The copyright is held by the author or creator of the new work. It is inherited as ordinary property. The information cannot be reproduced as such without written permission of the author or creator. However, the information or idea can be used by any one for creating the new work. The owner of a copyright can exclude others from reproducing, adapting, publicly distributing, publicly displaying or performing the copyrighted work.
8. **Enforcement:** The copyright comes into effect immediately after release of the matter covered by copyright such as book, movie, sound recording, *etc.*
9. **Rights Protected:** The purpose of copyright law is to encourage creative work by granting a temporary monopoly in an author's original creation. It controls the right of:
   (*i*) Reproduction of work *i.e.* copying,
   (*ii*) Creation of derivative work,
   (*iii*) Distribution of work,
   (*iv*) Performance of the work,
   (*iv*) Display of the work, and
   (*vi*) Digital transmission of performance.

   The law of copyright protects the first two rights in both private and public contexts, whereas an author can only restrict the last four rights in the public sphere.
10. **Infringement:** Infringement of the copyrighted material is possible. Because it an unregistered right. The official action cannot be taken against infringement of unregistered copyrights. The following acts by a person other than the author amount to infringement.
    (*i*) Reproducing or copying of the work
    (*ii*) Creating a new work derived from the original work such translating the work into a new language and transferring the work into a new medium of expression.
    (*iii*) Sell or give away the work or a copy of the work. Before transfer of copyright to that person.
    (*iv*) Performing or displaying the work in public without permission from the copyright

11. There are two basic requirements for protection of work under copyright *viz.*, (1)originality and (2) fixed expression.

12. **Literary Work:** It includes books [novels, instruction manuals etc], computer programs, lyrics for songs, articles in news papers and some types of database.

13. **Dramatic Work:** It includes works of dance or mime.

14. **Artistic Work:** It includes graphics and pictures such as paintings, engravings, photographs, sculptures, collages, works of architecture, technical drawings, diagrams, maps, logos *etc.*

15. **Sound Recordings:** It includes recordings on tape and compact disc. It also includes recordings of other copyright work such as musical or literary.

16. **Musical and choreographic work**

17. **Audiovisual Work:** Such as Movies andvideos; and

18. **Broad casting work of various types.**

19. **Work for Hire:** A work which is created by an employee in the regular course of his employment, it is considered a **work for hire** and the employer, not the employee, is considered the author of the work for copyright purposes.

C. Multiple Choice Questions

1. **Copyright is applicable to:**

   (a) Books (b) Movies

   (c) Music (d) All of these

2. **Copyright is applicable to:**

   (a) Paintings (b) Phonographs

   (c) Software (d) All of these

3. **Copyright is applicable to:**

   (a) Paintings (b) Engravings

   (a) Photographs (d) All of these

4. **Copyright is not applicable to:**

   (a) Ideas (b) Process

   (c) Procedures (d) All of these

5. **Copyright is not applicable to:**
   (a) Systems (b) Concepts
   (c) Discoveries (d) All of these

6. **In India, duration of Copyright is:**
   (a) 40 Years (b) 50 Years
   (c) 60 Years (d) 70 Years

7. **Related to copyrights, Literary work includes:**
   (a) Books (b) Lyrics
   (c) Computer programs (d) All of these

8. **Copyright is applicable to:**
   (a) Literary work (b) Dramatic
   (c) Artistic work (d) All of these

9. **Copyrighted work can be used for:**
   (a) Education (b) Book writing
   (c) Parody (d) All of these

10. **Copyright owner prevents others from:**
   (a) Reproducing (b) Adapting
   (c) Distribution (d) All of these

D. True or False

1. Copyright is a form of intellectual property right.
2. Copyright come under primary intellectual property rights.
3. In India, copyright provides protection for 60 years after the death of author.
4. Registration of copyright is not compulsory.
5. Copyright is held by the author or creator of the work.
6. Copyright comes in to force immediately after release of the work.
7. Copyright is valid in all countries of the world.
8. The author may transfer the copyright to any other party if he so desires.
9. The fair use of copyrighted work is permitted.
10. Infringement of copyrighted work is possible.
11. Copyright is inherited as ordinary right.

12. Literary work is protected by copyrights.
13. Artistic work comes under copyright.
14. Dramatic work comes under copyright.
16. Audiovisual work comes under copyrights.
17. Paintings are covered by copyrights.
18. Movies come under copyright.
19. Phonographs come under copyright.
20. Broadcasting comes under copyrights.

E. Fill in the Blanks

1. Copyright is a form of ________ property right.
2. Copyright come under ________ intellectual property rights.
3. In India, copyright provides protection for ________ years after the death of author.
4. Registration of copyright is not ________.
5. Copyright is held by the author or creator of the ________.
6. Copyright comes in to force immediately after ________ of the work.
7. Copyright is valid in ________ countries of the world.
8. Transfer the copyrighted matter to any other party ________.
9. Fair use of copyrighted work is ________.
10. Infringement of copyrighted work is ________.
11. Copyright is inherited as ________ right.
12. Literary work is protected by ________.
13. Artistic work comes under ________.
14. Dramatic work comes under ________.
16. Audiovisual work comes under ________.
17. Paintings are covered by ________.
18. Movies come under ________.

19. Phonographs come under ________

20. Broadcasting comes under ________.

F. Answer in Short or One Word

1. The exclusive rights granted to author/creator to reproduce, sell and distribute a work, prepare derivative works and display the work publicly.

2. An unauthorized use of material protected by copyright, patent or trademark law.

3. The privilege to use the copyrighted material in a reasonable manner without the consent of the owner.

4. A work which is created by an employee in the regular course of his employment.

5. Copying another's words.

ANSWERS

C. Multiple Choice Questions

Sl.No.	*Answer*	*Sl.No.*	*Answer*
1.	(d) All of these	**6.**	(c) 60 Years
2.	(d) All of these	**7.**	(d) All of these
3.	(d) All of these	**8.**	(d) All of these
4.	(d) All of these	**9.**	(d) All of these
5.	(d) All of these	**10.**	(d) All of these

D. True or False

1 to 20 all are true

E. Fill in the Blanks

Sl.No.	*Answer*	*Sl.No.*	*Answer*
1.	Intellectual	**11.**	Ordinary
2.	Primary	**12.**	Copyright
3.	60	**13.**	Copyright
4.	Compulsory	**14.**	Copyright
5.	Work	**15.**	Copyright
6.	Registration	**16.**	Copyright
7.	All	**17.**	Copyright
8.	Permitted	**18.**	Copyright
9.	Permitted	**19.**	Copyright
10.	Possible	**20.**	Copyright

F. One Word Answer

Sl.No.	*Answer*	*Sl.No.*	*Answer*
1.	Copyright	**2.**	Infringement
3.	Fair use	**4.**	Hire work
5.	Plagiarism		

CHAPTER 3

Patent Rights

A. Definitions

Business Method Patent: A specific type of software patent on a computer way of transacting business.

Design Patent: A type of patent that is granted to visual ornamental characteristics such as shape and surface of an article of manufacture.

Full-term Patent: Patents that last for a period of 20 years.

License: A promise by the patent owner (the licensor) not to sue the licensee for exercising on of the patent owner's righrs.

Patent: A patent is a set of exclusive rights granted by a state to a patentee (the inventor or assignee) for a fixed period of time.

Patent infringement: An un authorized use of patented invention/innovations.

Patent prosecution: The legal process of obtaining a patent from a patent office.

Plant Patent: A type of patent that is granted to newly developed plants having novel traits and fulfilling certain requirements.

Pre- disclosure: Disclosure of an invention before patenting; also known as on sale.

Prior art: An invention disclosed publicly. It can be cited against a new patent application.

Provisional Patent Application: An application filed before filing regular patent application to establish a filing date which may later be claimed as priority date.

Short Term Patent: Patents that last for a maximum period of 10 years.

Utility Patent: A type of patent that is granted to inventions that perform useful functions.

B. Background Information

1. Patent is a legal document issued by the patent office that grants exclusive rights for the production, sale and profit from the invention of a product or process for a specific period of time. Patents also grant the right to prevent others from copying the invention.

2. The main purpose of patent law is to promote the progress of science and useful arts by providing exclusive rights to inventors for their inventions for a limited period so that inventors can get advantage of their inventions.

3. Patents have been classified in various ways. On the basis of duration, patents are of two types, *viz.* (1) full-term patents and (2) short-term patents.

4. The maximum period for grant of a patent is 20 years. Those patents that are granted for a period of 20 years are referred to as full-term patents.

5. For the grant of full-term patent, the applicant must provide evidence of the novelty of invention. Thus full-term patents have two main features as follows.

6. Those patents that are granted for less than 20 years are called short-term patents. Such patents are granted for smaller inventors.

7. The patents are also classified as (i) utility patents, (ii) design patents, and (iii) plant patents. These are briefly discussed as follows:

8. A type of patent that is granted to inventions that perform useful functions is referred to as utility patent. This is also known as patent for invention.

9. A type of patent that is granted to visual ornamental characteristics such as shape and surface of an article of manufacture is called design patent. It is also called as picture patent. It is granted for a period of 14 years.

10. **Plant Patent:** A type of patent that is granted to new or newly developed, distinct plants reproducing asexually.

11. A patent provides protection for a specific duration. The period of protection varies from 15 to 20 years. In India it is 7-14 years. The long term patents are protected for 20 years in many countries and short term patents for 14 years.

12. The registration is compulsory for the patent. In other words, patent is a registered right. An application is submitted in prescribed form to the patent office for the registration of the paten and some fee is also paid for registration.

13. **Validity:** The patent is valid in the country where it has been granted. In future the patent granted by WTO will be valid in all member countries.

14. **Transfer:** The inventor has the power to transfer the right of patent to any other person if he so desires. In other words, the inventor may transfer ownership to any one [including a corporation]. Employee often assign the rights of their invention to their employers as part of their employment contracts.

15. **Items Covered:** Patent is applicable to new process, products, invention, device and innovations which have commercial or industrial applications.

16. **Control:** The inventor of the process, product or invention has control over the patent Patents provide exclusive rights to the inventor for the production, sale and profit from his invention or a product or process for a specific period of time.

17. **Enforcement:** The patent becomes effective immediately after its registration.

18. **Infringement:** Infringement of the patent refers to an unauthorized use of the process or product by other person. A patent owner may recover money damages in the form of a reasonable royalty. If the infringement is found willful, three times money is recovered.

19. **Rights Protected:** A utility patent once granted, provide the owner of the patent the right to prevent others from making, using, selling and importing a product covered by the utility patent.

20. **Requirements for Patents:** Four criteria, *viz.*, patentable subject matter, novelty, usefulness and non-obviousness are sufficient for granting patent

21. The patentable subject matter include five main items, *viz.*, (1) processes, (2) machines (3) articles of manufacture, (4) business methods and (5) software related ideas. Besides, new algorithms, data structure, new applications, business procedure and new technology can be patented. The same software may qualify for both copyright and patent protection, if it satisfies requirements for both of these.

22. **Non-patentable Matters include** (i) Ideas, mental processes, laws of nature and physical phenomena. For example one cannot patent a new mineral, insect or plant discovered or found in the wild. (ii) Literary, dramatic, musical and artistic work. These are subject to copyright protection. (iii) Inventions that are not useful or that are offensive to public morality.

23. An inventor cannot patent an inventions if he or she discloses it to the public more than one year before filing application for patent prosecution. Pre-disclosure is known as on sale which bar to patentability. The legal process of obtaining a patent from a patent office is called patent prosecution. An application for a patent is prepared by a professional patent agent or patent attorney and filed with a patent office where it is

examined by a patent examiner for patenting. Patent protection in multiple countries requires filing of separate patent applications in each country.

24. Four criteria, *viz.*, patentable subject matter, novelty, non-obviousness and sufficient disclosure are sufficient fro granting patent.

C. Multiple Choice Questions

1. **Full term patent is protected for a period of:**

   (a) 5 Years (b) 10 Years

   (c) 15 Years (d) 20 Years

2. **Short term patent is protected for a period of:**

   (a) 5 Years (b) 14 Years

   (c) 15 Years (d) 20 Years

3. **A utility patent is protected for a period of:**

   (a) 5 Years (b) 10 Years

   (c) 15 Years (d) 20 Years

4. **Which of the following is protected for 20 years?**

   (a) Copyrights (b) Full term patent

   (c) Short term patent (d) Trademark

5. **Which of the following is protected for 14 years?**

   (a) Copyrights (b) Full term patent

   (c) Short term patent (d) Trademark

6. **A patent is applicable to a new:**

   (a) Process (b) Invention

   (c) Product (d) All of these

7. **Basic requirement for a patent include:**

   (a) Novelty (b) Usefulness

   (c) Non-obviousness (d) All of these

8. **Non-patentable items refers to:**

   (a) Book (b) Process

   (c) Invention (d) Machine

9. **Non patentable item includes:**

   (a) Literary (b) Dramatic

   (c) Music (d) All of these

10. The patent holder has rights for:

(a) Production (b) Sale

(c) Authorization (d) All of these

11. A patent becomes effective immediately after its:

(a) Publication (b) Disclosure

(c) Registration (d) None of these

12. Non patentable item includes:

(a) Photograph (b) Invention

(c) Process (d) Machine

D. True or False

1. Long term patents are granted for 20 years.
2. Short term patents are granted for 14 years.
3. Utility patents are granted for 20 years.
4. A design patent is also called a picture patent.
5. Registration is compulsory for a patent.
6. The patent is valid in the country where it has been granted.
7. Utility patent is related to inventions.
8. The inventor can transfer the right of patent to any other person.
9. The patent becomes effective immediately after its registration.
10. A design patent is also known as picture patent.
11. Design patent is granted for 20 years.
12. A patent can also be granted to new plant variety.
13. Items which occur in nature cannot be patented.
14. A new item can be patented within one year of its invention.
15. An idea cannot be patented.
16. Inventions which are not useful for public cannot be patented.
17. A harmful invention cannot be patented.
18. Disclosure of an invention before patenting is called pre-disclosure.

19. Useful patents refer to utility patents.
20. Pre-disclosure is also known as sale.

E. Fill in the Blanks

1. Long term patents are granted for ________ years.
2. Short term patents are granted for ________ years.
3. Utility patents are granted for ________ years.
4. A design patent is also called a ________ patent.
5. For a patent ________ is compulsory.
6. The patent is valid only in the country of its ________.
7. Utility patent is related to ________
8. The inventor can ________ the right of patent to any other person.
9. The patent becomes effective immediately after its ________.
10. A design patent is also known as ________ patent.
11. Design patent is granted for ________ years.
12. Plant patent can be granted to a ________ plant variety.
13. Items which occur in ________ cannot be patented.
14. A new item can be patented within ________ of its invention.
15. An idea ________ be patented.
16. Inventions which are not useful for ________ cannot be patented.
17. A ________ invention cannot be patented.
18. Disclosure of an invention before patenting is called ________.
19. Useful patents refer to ________ patents.
20. Pre-disclosure is also known as ________.

F. Answer in Short or One Word

1. A set of exclusive rights granted by a state to a patentee (the inventor or assignee) for a fixed period of time.
2. An unauthorized use of patented invention/innovations.
3. Patents that last for a period of 20 years.

4. Patents that last for a maximum period of 10 years.
5. A type of patent that is granted to inventions that perform useful functions.
6. A type of patent that is granted to visual ornamental characteristics such as shape and surface of an article of manufacture.
7. A type of patent that is granted to new plant varieties having novel traits.
8. The legal process of obtaining a patent from a patent office.
9. Disclosure of an invention before patenting.
10. Publicly disclosed invention before patenting.
11. A specific type of software patent on a computer way of transacting business.
12. A promise by the patent owner (the licensor) not to sue the licensee for exercising the patent owner's rights.
13. An application filed before filing regular patent application to establish a filing date which may later be claimed as priority date.

ANSWERS

C. Multiple Choice Questions

Sl.No.	*Answer*	*Sl.No.*	*Answer*
1.	(d) 20 Years	**7.**	(d) All of these
2.	(b) 14 Years	**8.**	(a) Book
3.	(d) 20 Years	**9.**	(d) All of these
4.	(b) Full term patent	**10.**	(d) All of these
5.	(c) Short term patent	**11.**	(c) Registration
6.	(d) All of these	**12.**	(a) Photograph

D. True or False

1 to 20 all are true

E. Fill in the Blanks

Sl.No.	*Answer*	*Sl.No.*	*Answer*
1.	20	**11.**	20
2.	14	**12.**	New
3.	20	**13.**	Nature
4.	Picture	**14.**	1 Year
5.	Registration	**15.**	Cannot
6.	Registration	**16.**	Public
7.	Invention	**17.**	Harmful
8.	Transfer	**18.**	Pre-disclosure
9.	Registration	**19.**	Utility
10.	Picture	**20.**	Sale

F. Short or one word Answer

Sl.No.	*Answer*	*Sl.No.*	*Answer*
1.	Patent	**8.**	Patent prosecution
2.	Patent infringement	**9.**	Pre-disclosure
3.	Full term patent	**10.**	Prior art
4.	Short term patent	**11.**	Business method patent
5.	Utility patent	**12.**	License
6.	Design patent	**13.**	Provisional patent application
7.	Plant patent		

CHAPTER 4

Trademark Rights

A. Definitions

Arbitrary Mark: An arbitrary mark is one that bears no logical relationship to the underlying product.

Certification Marks: A certification mark is a symbol, name or device used by an organization to guarantee for the quality products and services provided by others.

Collective Mark: A trademark used by members of a group or organization is known as collective mark. It may be a symbol, label, phrase or other mark.

Descriptive mark: A descriptive mark directly describes rather than suggests, a characteristic or quality of underlying product such as its color, odor, function, dimensions or ingradients.

Generic Mark: A generic mark is a mark that describes the general category to which the underlying product belongs.

Service Marks: A trademark that is used in relation to services rather than products.

Suggestive Mark: This is a mark that evokes or suggests a characteristic of the underlying goods.

Trade Dress: A trademark that is known by its distinctive packaging is referred to as trade dress,.for example the yellow packaging of Kodak film.

Trademarks: Trademark is used to identify its product and to distinguish them from others. It is the name of a product made by a particular person or company.

Trade Search: A search that is conducted to know whether the mark is already in the use by some other party or not.

B. Background Information

1. A trademark is a distinctive sign of some kind which is used by an individual, business organization or other legal entity to uniquely identify the source of its products and/or services to consumers, and to distinguish its products or services from those of other entities.

2. **Period:** Initial term for a trademark registration is 10 years. It can last as long as the owner wants. Between the fifth and sixth year after the date of initial registration, the trademark registrant will need to file documents to keep the mark alive. A renewal then be filed every 10 years. Thus it can last indefinitely.

3. **Registration:** It is not essential to register a trademark. However, it is advisable to register the trademark to get legal rights against infringement of trademarks. Almost all countries in the world register and protect trademarks. The prescribed Fee has to be paid for registration of trademark.

4. **Validity:** The trademark rights are National, not International. Therefore, registration in one country does no confer rights in other countries. A separate application will be needed for registration in other country.

5. **Transfer:** The owner of the trademark has right to authorize another person to use it on payment (royalty) basis.

6. **Matters Covered:** A trademark includes a distinctive word, phrase, logo, domain name, graphic symbol, slogan, image or other device used to identify the source of a product from others. Thus a variety of items are covered by trademark. Recently colors [such as John deere green], sounds and scents have also been registered as trademarks.

7. **Control:** A trademark owner has rights to prevent unauthorized use of trademark.

8. **Enforcement:** The trademark becomes effective when it is first publicly put to continuous commercial use, even without registration.

9. **Infringement:** An unauthorized use of a trademark by a person other than the trademark owner is called infringement of trademark. Infringement occurs when someone uses the same or similar trademark for a similar good or service. Infringement does not require the infringer to be competitor, but infringement does require a similar product.

10. **Rights Protected:** Protection of trademarks provides exclusive right to the owner of the trademark to use it to identify goods or services.

11. **Requirements:** Trademarks apply to products and services. There are some requirements for registration of a trademark. It must be distinctive,

not descriptive and there must be no earlier rights or multiple appearance. Moreover, it should not confuse with other trademark. It should not be prohibitive (reserved) and generic.

12. There are many types of trademarks such as, (!) service mark, (2) trade dress, (3) collective mark and (4) certification mark. These are defined as follows:

13. **Service Mark:** A trademark that is used in relation to services rather than products is called a service mark. It promotes services or events. Some examples of service marks are as follows.

(*i*) Amazon.com [retail website],

(*ii*) B. Jack in the Box [fast food service],

(*iii*) Kinko's [photocopying service]

(*iv*) Blockbuster [video rental service] and

(*v*) Fed Ex logo delivery service].

14. **Trade Dress:** A trademark that is known by its distinctive packaging is referred to as trade dress, for example the yellow packaging of Kodak film. The trade dress often serves the same function as a trademark or service mark.

15. **Collective Mark:** A trademark used by members of a group or organization is known as collective mark. It may be a symbol, label, phrase or other mark. These are used to identify goods, products or services rendered by a group. Collective marks are often used to show membership in a union, association or other organization.

16. **Certification Mark:** A certification mark is a symbol, name or device used by an organization to guarantee for the quality products and services provided by others such as Good Housekeeping Seal of Approval.

Based on the relationship, between the mark and the underlying product, trademarks are of four types,*viz.* (1) arbitrary, (2) suggestive, (3) descriptive, and (4) generic. These are defined as under.

17. **Arbitrary Mark:** An arbitrary mark is one that bears no logical relationship to the underlying product. For example, the words Ëxxon"and "Kodak"bear no inherent connection to their underlying products. It is also known as fanciful mark.

18. **Suggestive Mark:** This is a mark that evokes or suggests a characteristic of the underlying good. For example the word "Copper tone" suggests the color of a deep sun tan.

19. **Descriptive Mark:** A descriptive mark directly describes rather than suggests, a characteristic or quality of underlying product such as its color, odor, function, diamensions or ingradients. For example "Holiday Inn"and A"ll Bran"describe hotel rooms and breakfast cereal.

20. **Generic Mark:** A generic mark is a mark that describes the general category to which the underlying product belongs For example, the term" Computer" is a generic term for computer equipment. Such terms are not entitled for protection.

PROCEDURE OF FILING TRADEMARKS

The application for registration of trademark is to be submitted in the Patent and Trademark Office. The prior use of the trademark for commercial use is required for registration. The registration process consists of the following steps:

1. **Filling of Application:** The prescribed application form is used and the application has to be carefully completed and signed. The following documents should be attached with the application. Attached to the goods, containers for the goods.
   (*i*) Drawing of the mark on separate piece of paper.
   (*ii*) **Prescribed Fee:** The Fee is not refundable even if the application is rejected.
   (*iii*) **Specimens:** The manner in which the mark is to be used should be indicated. The application must clear reproduction of the sign filed for registration including, any colors, forms or three. dimensional features. The application must also contain a list of goods or services to which the sign would apply. The sign must be distinctive to identify the product from other products. Each application must include three specimens.
2. **Examination of the application:** The application is examined in the PTO by an examining attorney. If the application is found correct, the attorney will approve the mark for publication in the official Gazette. A trademark search may also be conducted to know whether the mark is already in the use by some other party.
3. **Opposition:** If any party thinks that the mark will damage it rights, opposition can be filed within 30 days from the date of publication in the Gazette.
4. **Registration:** If no opposition is filed, the PTO will register the trademark and issue a registration certification a reasonable time.

C. Multiple Choice Questions

1. **Initially trademark is protected for a period of:**

(a) 5 Years (b) 10 Years

(c) 15 Years (d) 20 Years

2. **A trademark includes a distinctive:**
   (a) Word (b) Phrase
   (c) Logo (d) All of these

3. **A trademark includes a distinctive:**
   (a) Graphic symbol (b) Slogan
   (c) Image (d) All of these

4. **Trade dress is a type of:**
   (a) Copyright (b) Trademark
   (c) Patent (d) All of these

5. **A certification mark may be a:**
   (a) Symbol (b) Name
   (c) Device (d) All of these

6. **A trademark comes under:**
   (a) Primary rights (b) Secondary rights
   (c) *Sui-generis* rights (d) none of these

7. **After publication in a gazette, opposition can be filed within:**
   (a) 30 days (b) 40 days
   (c) 45 days (d) 60 days

8. **Trademark rights are:**
   (a) Regional (b) National
   (c) Local (d) International

9. **A descriptive trademark refers to:**
   (a) Color (b) Odor
   (c) Function (d) All of these

10. **A descriptive trademark refers to:**
   (a) Dimensions (b) Ingredient
   (c) Function (d) All of these

D. True or False

1. A trademark is a primary intellectual property right.
2. For a trademark registration is not compulsory.
3. Registration provides rights against infringement.
4. Initially trademark can be registered for 10 years.

5. Trademarks are related to products and services.
6. Trademark rights are National.
7. Transfer of trademark is possible.
8. A trademark used for services is called service mark.
9. A trademark used for distinctive packaging is called trade dress.
10. A trademark becomes effective when it is first put for public use.
11. Trademarks are renewable indefinitely.
12. A trademark used by an organization refers to collective mark.

E. Fill in the Blanks

1. A trademark is a ________ intellectual property right.
2. For a trademark registration is ________ compulsory.
3. Registration provides rights against ________.
4. Initially trademark can be registered for ________ years.
5. Trademarks are related to ________.
6. Trademark rights are ________.
7. Transfer of trademark is ________.
8. A trademark used for services is called ________.
9. A trademark used for distinctive packaging is called ________.
10. A trademark becomes effective when it is first put for ________ use.
11. Trademarks are renewable ________.
12. A trademark used by an organization refers to ________.

F. Answer in Short or One Word

1. A mark that bears no logical relationship to the underlying product.
2. A mark that evokes or suggests a characteristic of the underlying goods.
3. A mark that directly describes a characteristic or quality of underlying product such as its color, odor, function, dimensions or ingredient.
4. A mark that describes the general category to which the underlying product belongs.

5. A mark which is used to identify its product and to distinguish them from others.

6. A trademark that is used in relation to services rather than products.

7. A certification mark is a symbol, name or device used by an organization to guarantee for the quality products and services provided by others.

8. A trademark that is known by its distinctive packaging is referred to as trade dress, for example the yellow packaging of Kodak film

9. A trademark used by members of a group or organization.

10. A search that is conducted to know whether the mark is already in the use by some other party or not.

ANSWERS

C. Multiple Choice Questions

Sl.No.	Answer	Sl.No.	Answer
1.	(b) 10 Years	6.	(a) Primary rights
2.	(d) All of these	7.	(a) 30 days
3.	(d) All of these	8.	(b) National
4.	(d) All of these	9.	(d) All of these
5.	(d) All of these	10.	(d) All of these

D. True or False

1 to 12 all are true

E. Fill in the Blanks

Sl.No.	Answer	Sl.No.	Answer
1.	Primary	7.	Possible
2.	Not	8.	Service mark
3.	Infringement	9.	Trade dress
4.	10	10.	Public
5.	Product and services	11.	Indefinitely
6.	National	12.	Collective mark

F. Short or One Word Answer

Sl.No.	Answer	Sl.No.	Answer
1.	Arbitrary mark	6.	Service mark
2.	Suggestive mark	7.	Certification mark
3.	Descriptive mark	8.	Trade dress
4.	Generic mark	9.	Collective mark
5.	Trade mark	10.	Trade search

Chapter 5

Industrial Design

A. Definitions

Industrial Design: The features of shape, configuration, pattern or ornament [or any combination thereof] applied to a finished article.

Industrial Design Protection: A form of monopoly like other types of intellectual property.

Protected Design: A design registered with Patents and Trademarks Office.

B. Background Information

1. An industrial design refers to any original shape, picture, or some combination applied to a useful article of manufacture.

2. Novelty and originality are important criteria in a design for registration. In addition, only those designs that are applied to an article by an industrial process are protected.

3. **Period:** Initially the design is registered for a term of 5 yearly basis. Later on it can be renewed on 5 years by paying the prescribed Fee. In some cases, it may be registered up to 15 years.

4. **Registration:** Registration is essential to get legal rights. There is no protection without registration. A nominal processing Fee is charged for registration.

5. **Validity:** The protection of the design is valid only in the country where it has been registered. The protection in other countries can be obtained by filling separate application in each country.

6. **Transfer:** A registered design is a property like any other commodity which can be purchased, sold, hired or licensed. It is advisable to register

the sale of rights in the registration office before use of the design by new owner.

7. **Matters Covered:** Industrial designs include any original shape, picture, or some combination applied to a useful article of manufacture. Industrial designs do not cover following things.
   (*i*) The functional features of an article,
   (*ii*) A principle of construction, or how an article is built,
   (*iii*) The materials used in the construction of an article
   (*iv*) Color per se, and Ideas
8. **Control:** The owner has the legal rights to prevent others from unauthorized use of his design.
9. **Enforcement:** The industrial design comes into force immediately after it registration in the patents and trademarks office.
10. **Infringement:** Unauthorized use of a design is called infringement of such design. The owner has the right to take legal action against the infringer and claim damages. However, the legal action must be taken within three years of the alleged offence.
11. **Rights Protected:** It provides exclusive rights to the owner such as to make,imports, sell or hire out or allow others to use the design under terms settled with the registered owner.
12. **Requirements:** There are two requirements of industrial design for registration, *viz.*, novelty and commercial application. In other words, it should different from previously registered designs and it should not be made public before registration. It is made open to public after registration.

Procedure for Registration

The registration is done in Patents and trademarks Office. The owner of the design has to complete the following formalities:

1. He has to submit duly filled up application in prescribed form in the Patents Office. Prescribed Fee has to paid along with application
2. The specimen of the design is to be provided. The only visual aspects of the design such as shape, configuration, pattern or ornament or any combination thereof should be described. It must state clearly what the design is. Too much details are not required.
3. The registry office will make a search about novelty of the design.
4. If requirements are met, a certificate of registration is issued otherwise registration will be refused which is appeal able.
5. Initially 5 years registration is granted that can be renewed on five yearly basis.

6. Normally, the registration process should be completed within six months. However, a total of 12 months [extendable to 15 months] time is allowed for completion of registration process. The design should not be made public before registration otherwise registration may later be proved invalid.

C. Multiple Choice Questions

1. Initial period of registration for an industrial design is:

(a) 2 Years (b) 3 Years
(c) 5 Years (d) 7 Years

2. Industrial design can be renewed after every:

(a) 5 Years (b) 7 Years
(c) 10 Years (d) 14 Years

3. An industrial design includes:

(a) Shape (b) Ornament
(c) Pattern (d) All of these

4. An industrial design makes the product:

(a) Aesthetic (b) attractive
(c) Decorative (d) All of these

5. An industrial design can be:

(a) Purchased (b) Sold
(c) Licensed (d) All of these

6. Owner of an industrial design has the right of:

(a) Selling (b) Exporting
(c) Authorization (d) All of these

7. Normally, registration of an industrial design is completed in:

(a) 3 months (b) 4 months
(c) 6 months (d) 1 year

8. In India, industrial design Act was introduced in:

(a) 1960 (b) 1990
(c) 2000 (d) 2001

D. True or False

1. Industrial design comes under primary intellectual property rights.
2. Initial period of registration is 5 years.

3. Later on it can be renewed after every 5 years.
4. Registration of Industrial design is essential to get legal rights.
5. Industrial design enhances aesthetic value of a product.
6. Industrial designs are liable to infringement.
7. Originality and novelty are essential for registration of Industrial designs.
8. Industrial designs come into force immediately after registration.
9. Industrial designs are valid only in the county of registration.
10. Rights of Industrial design can be transferred to any other person.
11. Industrial design is judged by eye inspection only.
12. Industrial designs may lead to monopoly.

E. Fill in the Blanks

1. Industrial design comes under ________ intellectual property rights.
2. Initial period of registration is ________ years.
3. Later on it can be renewed after every ________ years.
4. Registration of Industrial design is ________ to get legal rights.
5. Industrial design enhances ________ value of a product.
6. Industrial designs are liable to ________.
7. Originality and novelty are ________ for registration of Industrial designs.
8. Industrial designs come into force immediately after ________.
9. Industrial designs are valid only in the county of ________.
10. ________ of rights of Industrial design to any other person is possible.
11. Industrial design is judged by ________ only.
12. Industrial designs may lead to ________.

F. Answer in Short or One Word

1. The features of shape, configuration, pattern or ornament applied to a finished article.
2. A design registered with Patents and Trademarks Office.
3. A form of monopoly like other types of intellectual property.

ANSWERS

C. Multiple Choice Questions

Sl.No.	Answer	Sl.No.	Answer
1.	(c) 5 Years	5.	(d) All of these
2.	(a) 5 Years	6.	(d) All of these
3.	(d) All of these	7.	(c) 6 months
4.	(d) All of these	8.	(c) 2000

D. True or False

1 to 12 all are true

E. Fill in the Blanks

Sl.No.	Answer	Sl.No.	Answer
1.	Primary	7.	Essential
2.	5	8.	Registration
3.	5	9.	Registration
4.	Essential	10.	Transfer
5.	Aesthetic	11.	Eye inspection
6.	Infringement	12.	Monopoly

F. Short or One Word Answer

Sl.No.	Answer	Sl.No.	Answer
1.	Industrial design	2.	Protected design
3.	Industrial design protection		

Chapter 6

Utility Models

A. Definitions

Utility Model: A utility model is an exclusive right granted for an invention, which allows the right holder to prevent others from commercially using the protected invention, without his authorization, for a limited period of time.

Utility Model Infringement: Unauthorized use of a utility model by any person or company.

B. Background Information

1. A utility model is an exclusive right granted for an invention, which allows the right holder to prevent others from commercially using the protected invention, without his authorization, for a limited period of time.
2. Utility models are also known as petty patents, petty inventions, petty innovations, utility innovations, minor patents and small patent.
3. The rights conferred by utility models are similar to those granted by patent laws but has a shorter term.
4. They are registered with National Patent Office.
5. Presently, utility models are protected in about 170 countries of the World.
6. In some countries, utility model protection can only be obtained for certain fields of technology and only for products but not for processes.
7. Utility model means any form, configuration or disposition of elements of some appliance, working tools and implements as articles of everyday use, electrical and electronic circuitry of other object or part thereof in so far as they are capable of contributing some benefit or new effect of saving in time, energy, or labour or improving the hygienic or socio-physiological

working conditions by means of new configuration, arrangement or device or a combination thereof and are industrially applicable.

8. **Countries Where Protected:** A utility model is an intellectual property right to protect inventions. This right is available in more than 50 countries. It is very similar to the patent, but usually has a shorter term (often 6 or 10 years) and less stringent patentability requirements.

9. **Requirements for Registration:** Most countries having utility model laws require that the invention be new. However, many patent or utility model offices do not conduct substantive examination and merely grant the utility model after checking that utility model applications comply with formalities. The protection for utility models shall be available for utility models which are new and have industrial applications. Each utility model application shall be subjected to substantive examination for compliance with substantive examination requirements. Each designated state prior to registration shall be accorded the opportunity to give an opinion relating to registration in its territory of the utility model.

10. **Registration:** For the purposes of novelty and priority disclosures, utility model applications shall not be refused on grounds that the invention has been exhibited at an officially recognized exhibition before the date of filing. This exception will only apply if the exhibition occurred not more than six months before the date of filing.

    (*i*) **Filing Application:** Utility model applications may be prepared and filed at local patent offices in countries where utility model protection is available. The application for a utility model should contain (i) name of the applicant, (ii) description of the utility model, (iii) drawings of the model, (iv) abstract, (v) state for which utility model is to be protected, and (vi) prescribed fee.

    Conversion of application: Application refused by the office or application which have been refused consideration can on request be converted into national applications. Patent application can also be converted into a utility model application or vice versa on request and such a converted application can benefit from the filing date of the initial application.

    (*ii*) **Examination of the application:** The application is examined by the experts. There is a pre-registration examination for Utility Models. This is conducted in three stages. The first stage involves examination to determine whether the application meets the formalities requirements given under 4 above.

    The Second Stage involves substantive examination. The substantive examination is carried out to determine whether the utility model is industrially applicable and secondly whether it is new. Novelty in this case shall be the local Novelty in a designated state. On completion of the second stage, the application is forwarded to designated states.

The Third Stage involves examination by a designated state for novelty and for conformity with the national laws. This examination or verification should be completed within six months from the date of notification.

Each designated state is required to carry an examination to ascertain whether the Utility Model is New in the territory of the state designated.

(*iii*) **Registration and Publication:** Registration and Publication entails recording of the Utility Model in the Journal and the Register. It also entails issuance of a Certificate and Publication of a copy of the specification.

11. **Registration Fee:** The registration fee and maintenance charges are much lower than those for patents.

12. **Rights Covered:** Under the Utility Model Act, "the owner of a utility model shall have an exclusive right to commercially work the registered model" and is allowed to demand an injunction, compensation for damage, and compensation for unfair profits gained by the infringer.

13. **Authorization:** The inventor has the right to authorize another person to make use of his invention on royalty basis.

14. **Duration:** The utility model is registered for a shorter period than patent. The period of protection generally varies from 6 to 10 years in different countries.

15. **Validity:** The utility model will be valid in the country where it has been registered Provided The applicable national laws shall also apply.

16. **Infringement:** Infringement of utility model is possible. The owner or the inventor has legal rights to sue the infringer and recover damaged caused by way of infringement.

C. Multiple Choice Questions

1. **Utility models are protected for a period of:**

   (a) 5 Years (b) 6-10 Years

   (c) 12 Years (d) 15 Years

2. **Utility models include:**

   (a) Appliances (b) Working tools

   (c) Implements (d) All of these

3. **Utility models are also known as:**

   (a) Petty patents (b) Petty inventions

   (c) Petty innovations (d) All of these

4. **Utility models are also known as:**
   (a) Utility innovations (b) Minor patents
   (c) Small patents (d) All of these
5. **Novelty is much lower in:**
   (a) Utility patents (b) Patents
   (c) Copyrights (d) Trademark
6. **Utility models are related to:**
   (a) Invention (b) Formula
   (c) Ideas (d) All of these

D. True or False

1. Utility models come under primary intellectual property rights.
2. Utility models are also known as petty patents.
3. The utility model will be valid in the country where it has been registered.
4. The authorization of utility models to another person is permitted.
5. Utility models can be protected for 6-10 years.
6. Registration of utility models is easier than patents.
7. Utility model rights are granted for an invention.
8. A utility model can also be registered as a patent.
9. Registration fee for utility model is cheaper than a patent.
10. Infringement of utility model is possible.

E. Fill in the Blanks

1. Utility models come under ________ intellectual property rights.
2. Utility models are also known as ________ patents.
3. The utility model is valid in the country where it has been ________.
4. The ________ of utility models to another person is permitted.
5. Utility models can be protected for ________ years.
6. Registration of utility models is ________ ________ than patents.
7. Utility model rights are granted for an ________.
8. A utility model can also be registered as a ________.

9. Registration fee for utility model is ________ than a patent.

10. Infringement of utility model is ________.

F. Answer in Short or One Word

1. An exclusive right granted for an invention, which allows the right holder to prevent others from commercially using the protected invention, without his authorization, for a limited period of time.

2. Unauthorized use of a utility model by any person or company.

ANSWERS

C. Multiple Choice Questions

Sl.No.	*Answer*	*Sl.No.*	*Answer*
1.	(b) 6-10 Years	**4.**	(d) All of these
2.	(d) All of these	**5.**	(a) Utility patents
3.	(d) All of these	**6.**	(a) Invention

D. True or False

1 to 10 all are true

E. Fill in the Blanks

Sl.No.	*Answer*	*Sl.No.*	*Answer*
1.	Primary	**6.**	Easier
2.	Petty patents	**7.**	Invention
3.	Registered	**8.**	Patent
4.	Transfer	**9.**	Cheaper
5.	6-10	**10.**	Possible

F. Short or One Word Answer

Sl.No.	*Answer*	*Sl.No.*	*Answer*
1	Utility model	2	Utility model infringement

CHAPTER 7

Geographical Indication

A. Definitions

Geographical Indication: A sign used on goods that have a specific geographical origin and often possess qualities or a reputation that are due to that place of origin.

Infringement of GI: Unauthorized use of a geographical indication by another person

Agricultural GI: Reputation of a goods related to agriculture such as Basmati Rice, Nagpur Orange and Alphanso mango.

Natural GI: A reputed product occurring in nature such as Darjeeling Tea

Manufactured GI: Commercially produced reputed goods such as Kanchipuram Silk Saree, Banarasi Silk Saree, Kolhapuri Chappal, Bikaneri Bhujia and Agra Petha

B. Background Information

1. Geographical Indication may be defined as the name of a region, a specific place or, in exceptional cases a country, used to describe an agricultural, natural or manufactured goods, product or a food stuff.

2. The World Intellectual Property Organization (WIPO) helps in protection of Geographical Indications [GI]. About 170 geographical indications have been registered. Now this number would be much more.

3. In India, the Parliament has passed the Geographical Indications of Goods (Registration and Protection) Act, 1999 [in December 1999]. This Act seeks to provide for the registration for better protection of geographical indications relating to goods in India. The Act would be administered by the Controller General of Patents, Design and Trade Marks who is

the Registrar of Geographical Indications. The Geographical Indications Registry is located at Chennai.

4. So far geographical indications have been registered in some advanced countries because they have developed the system of protecting GI. In the past, wines, spirits, cheeses, tobacco, which account for 88 per cent, have been registered as GI. Wines and spirits account for almost 71 per cent of all registrations.

5. In addition to wines and spirits, there is need for protection of agricultural products as GI. For this there is need to establish a multilateral system of identification and registration of GI. Geographical Indication may be an agricultural, natural or manufactured goods or product.

6. **Period:** The registration of a geographical indication is valid for a period of 10 years. It can be renewed from time to time for further period of 10 years each. If a registered geographical indication is not renewed it liable to be removed from the register.

7. **Registration:** The registration is not compulsory. The registration provides better legal protection to take action against infringement. In other words, it confers legal protection to geographical indications. And prevents unauthorized use of a registered geographical indication by others.

8. **Validity:** The registration of the geographical indication is valid in the country of its origin.

9. **Transfer:** A geographical indication cannot be sold, purchased, transferred, licensed, pledged, mortgaged or put to such other agreement.

10. **Matters Covered:** It is used to identify agricultural, natural and manufactured goods.

11. **Control:** An authorized user has the exclusive rights o use the geographical indication in relation to goods for which it is registered.

12. **Enforcement:** The protection of geographical indication becomes effective after its registration.

13. **Infringement:** Infringement of a geographical indication is possible. The registered proprietor and authorized users can initiate action against infringement

14. **Rights Protected:** The proprietor or authorized user has the rights to use the geographical indication and prevent other from using such material without permission.

15. **Requirements:** It originates from a definite geographical territory and should have a special quality or reputation or other characteristics. The main features are as given below:

(*i*) it is an indication.

(*ii*) It originates from a definite geographical territory.

(*iii*) It is used to identify an agricultural, natural or manufactured goods/ product.

(*iv*) The manufactured goods should be processed or processed in that territory.

(*v*) It should have a special quality or reputation or other characteristics.

16. The registration is done by the registrar of Geographical Indications. For registration any association of persons, producers, organization or authority established by or under the law can apply The applicant must represent the interest of the producers. The registration consists of following steps:

(i) **Filing Application:** The application in the prescribed form is filed to the registrar of Geographical Indications along with prescribed Fee and other details.

(*ii*) **Examination:** The application is examined by the registrar office by conducting search. The application is accepted, if it fulfils the requirements. Then it is advertised in the GI journal for objection if any.

(*iii*) **Acceptance:** If there is no objection, the application is accepted for grant of registration.

(*iv*) **Registration:** Then the GI is entered in the Register. The register of geographical indication is divided into two parts, *viz.* part A and part B. Part A consists of particulars relating to registered geographical indication and part B consists of particulars of the registered authorized users.

(*v*) **Issue of Certificate:** Finally a certificate is issued to the registered user. The registration of GI is for a period of 10 years which is renewable for further periods of 10 years each.

17. In India some geographical indications have been identified which are listed below:

(i) Basmati Rice; (ii) Darjeeling Tea; (iii) Kanchipuram Silk Saree; (iv) Alphanso Mango; (v) Nagpur Orange; (vi) Kolhapuri Chappal; (vii)Bikeneri Bhujia; (viii) Agra Petha; (ix) Banarasi Saree.

The above geographical indications can be classified into three groups, *viz.* (i) agricultural, (ii) natural,and (iii) manufactured.

18. **Agricultural GI:** This group includes Basmati Rice, Nagpur Orange and Alphanso mango.

19. **Natural GI:** This group includes Darjeeling Tea.

20. **Manufactured GI:** This group includes Kanchipuram Silk Saree, Banarasi Silk Saree, Kolhapuri Chappal, Bikaneri Bhujia and Agra Petha.

C. Multiple Choice Questions

1. **Protection of geographical indication is valid for:**
   (a) 5 Years (b) 8 Years
   (c) 10 Years (d) 12Years

2. **In India, geographical indication Act was introduced in:**
   (a) 1990 (b) 1999
   (c) 2001 (d) 2004

3. **Bansmati rice is an example of:**
   (a) Natural GI (b) Agricultural GI
   (c) Manufactured GI (d) None of these

4. **Banarsi silk saree is an example of:**
   (a) Natural GI (b) Agricultural GI
   (c) Manufactured GI (d) None of these

5. **Alphanso mango is an example of:**
   (a) Natural GI (b) Agricultural GI
   (c) Manufactured GI (d) None of these

6. **Kanchipuram silk saree is an example of:**
   (a) Natural GI (b) Agricultural GI
   (c) Manufactured GI (d) None of these

7. **Agra petha is an example of:**
   (a) Natural GI (b) Agricultural GI
   (c) Manufactured GI (d) None of these

8. **Darjeeling tea is an example of:**
   (a) Natural GI (b) Agricultural GI
   (c) Manufactured GI (d) None of these

9. **Kolhapuri Chhapal is an example of:**
   (a) Natural GI (b) Agricultural GI
   (c) Manufactured GI (d) None of these

10. **In India, office of geographical indications is located at:**
    (a) New Delhi (b) Chennai
    (c) Kolkata (d) Mumbai

D. True or False

1. Geographical indications come under primary intellectual property rights.
2. Bansmati rice is an example of agricultural geographical indication.
3. Banarsi saree is an example of manufactured geographical indication.
4. Alphanso mango is an example of agricultural geographical indication.
5. Darjeeling tea is an example of natural geographical indication.
6. Kachipuram silk saree is an example of manufactured geographical indication.
7. Kolhapuri Chappal is an example of manufactured geographical indication.
8. Bikaneri Bhujia is an example of manufactured geographical indication.
9. Registration of geographical indication is valid for 10 years.
10. Agra Petha is an example of manufactured geographical indication.
11. In India geographical indication registry is located at Chennai.
12. In India, geographical indication act came into force in 1999.

E. Fill in the Blanks

1. Geographical indications come under ________ intellectual property rights.
2. Bansmati rice is an example of ________ geographical indication.
3. Banarsi saree is an example of ________ geographical indication.
4. Alphanso mango is an example of ________ geographical indication.
5. Darjeeling tea is an example of ________ geographical indication.
6. Kachipuram silk saree is an example of ________ geographical indication.
7. Kolhapuri Chappal is an example of ________ geographical indication.
8. Bikaneri Bhujia is an example of ________ geographical indication.
9. Registration of geographical indication is valid for ________ years.
10. Agra Petha is an example of ________ geographical indication.
11. In India geographical indication registry is located at ________.
12. In India, geographical indication act came into force in ________.

F. Answer in Short or One Word

1. A sign used on goods that have a specific geographical origin and often possess qualities or a reputation that are due to that place of origin.
2. Unauthorized use of a geographical indication by another person
3. Reputed goods related to agriculture.
4. A reputed product occurring in nature.
5. Commercial reputed products.

ANSWERS

C. Multiple Choice Questions

Sl.No.	Answer	Sl.No.	Answer
1.	(c) 10 Years	**6.**	(c) Manufactured GI
2.	(b) 1999	**7.**	(c) Manufactured GI
3.	(b) Agricultural GI	**8.**	(a) Natural GI
4.	(c) Manufactured GI	**9.**	(c) Manufactured GI
5.	(b) Agricultural GI	**10.**	(b) Chennai

D. True or False

1 to 12 all are true

E. Fill in the Blanks

Sl.No.	Answer	Sl.No.	Answer
1.	Primary	**7.**	Manufactured
2.	Agricultural	**8.**	Manufactured
3.	Manufactured	**9.**	10
4.	Agricultural	**10.**	Manufactured
5.	Natural	**11.**	Chennai
6.	Manufactured	**12.**	1999

F. Short or One Word Answer

Sl.No.	Answer	Sl.No.	Answer
1.	Geographical indication	**4.**	Natural GI
2.	Infringement of GI	**5.**	Manufactured GI
3.	Agricultural GI		

CHAPTER 8

Trade Secret

A. Definitions

Copyrights: The exclusive rights to reproduce, sell and distribute a work, prepare derivative works and display the work publicly.

Domain Name: The names and words that companies designate for their registered Internet Web site addresses.

Patent: A document granting an inventor sole rights to an invention. It is an official document which grants sole rights to the inventor for manufacturing and marketing his product/process/invention to derive benefits.

Trademark: A trademark can be a word, name, symbol, device or mark which is used to identify and distinguish the goods or services of one company from goods or services of another.

Trade Name: The name under which a company conducts business, or by which its business, product/service are identified. It may or may not be registered as a trademark.

Trade Secret: A formula or process or device used in business that is not published or divulged which gives an advantage over competitors.

B. Background Information

1. A trade secret is anything [a formula, process, method, mechanism, tool, pattern or device] which the disclosing party desires to keep secret.

2. There is no specific period for trade secret. It may continue life long or for generations together.

3. There is no need of registration for trade secret. Trade secret can never be patented, because it does not publicize the knowledge.

4. It does not provide opportunity to others for improvement of innovation. It is not applicable to books, equipments, plant varieties, designs which are openly used.

5. **Period:** It can be protected for indefinite period. In other words, it can be protected so long as the idea/formula is kept secret.

6. **Registration:** There is no need of registration for a trade secret.

7. **Validity:** It is valid with the company concerned.

8. **Transfer:** The owner of the trade secret can reveal the secrecy to other person on payment basis. Trade secret is a property right Hence the owner of the trade secret can sell his trade secret to someone else. Moreover, the owner can license his trade secret to others on royalty basis.

9. **Matters Covered:** A trade secret may consist of following things:

   A formula for a chemical compound,

   A process of manufacturing, trating opr preserving materials,

   A pattern for a machine or other device, or

   A list of customers.

   Novelty and inventions are not requisite for a trade secret as they are not for patentability.

10. **Control:** The trade secret is controlled by the owner of the company. Trade secrets protect the application of ideas.

11. **Enforcement:** The trade secret becomes effective when the product is put to continuous commercial use without disclosure.

12. **Infringement:** The infringement of the trade secter is possible through reverse engineering.

13. **Rights Protected:** Protection of trade secret provide the benefit of production and marketing only to the owner of the trade secret. The owner of a trade secret has the right to prevent use or dissemination of the trade secret by any one who learned or derived the trade secret from the owner. The owner has no trade secret rights against any one who independently discovers the trade secret.

14. **Requirements:** Any thing confidential and of commercial use can be protected by trade secret.

15. A trade secret can be safeguarded in various ways such (i) by obtaining written agreement from the employees, who know about the trade secret, not to disclose the secret to anyone and (ii) by keeping the documents, containing information about trade secret, in the locking system The employer may recover damages against such a person [employee]

revealing the secret. The written agreements are of two types, *viz.* A) non-disclosure agreements,and (b) non-compete agreements.

C. Multiple Choice Questions

1. **Trade secret is applicable to:**

   (a) Formula (b) Process

   (c) List of customers (d) all of these

2. **Trade secret can be safeguarded for:**

   (a) 10 years (b) 15 years

   (c) 20 years (d) indefinite period

3. **Which of the following can be safeguarded indefinitely?**

   (a) Copyright (b) Patent

   (c) Trade secret (d) Trade name

4. **Which of the following is related to a chemical formula?**

   (a) Patent (b) Copyright

   (c) Trade name (d) Trade secret

5. **Which of the following is confidential?**

   (a) Trademark (b) Patent

   (c) Trade secret (d) Trade name

6. **A chemical formula comes under:**

   (a) Copyright (b) Patent

   (c) Trade secret (d) Trade name

D. True or False

1. Trade secret comes under primary intellectual property rights.
2. Registration is not required for trade secret.
3. Trade secret is valid in the company concerned.
4. Trade secret is controlled by owner of company.
5. Trade secret can be safeguarded indefinitely.
6. Trade secret is liable to infringement.
7. Trade secret is related to chemical formula.
8. Trade secret becomes effective when the product is put to commercial use.

9. Trade secret can be discovered by reverse engineering.
10. Rights of trade secret are vested with the owner.
11. Trade secret protects the application of ideas.

E. Fill in the Blanks

1. Trade secret comes under ________ intellectual property rights.
2. ________ is not required for trade secret.
3. Trade secret is valid in the ________ concerned.
4. Trade secret is controlled by ________ of company.
5. Trade secret can be safeguarded ________.
6. Trade secret is liable to ________.
7. Trade secret is related to ________ formula.
8. Trade secret becomes effective when the product is put to ________ use.
9. Trade secret can be discovered by ________.
10. Rights of trade secret are vested with the ________.
11. Trade secret protects the application of ________.

F. Answer in Short or One Word

1. A formula or process or device used in business that is not published or divulged which gives an advantage over competitors
2. The exclusive rights to writer or creators to reproduce, sell and distribute a work, prepare derivative works and display the work publicly.
3. A document granting an inventor sole rights for an invention.
4. A word, name, symbol, device or mark which is used to identify and distinguish the goods or services of one company from goods or services of another.
5. The name under which a company conducts business, or by which its business, product/service are identified.
6. The names and words that companies designate for their registered Internet Web site addresses.

ANSWERS

C. Multiple Choice Questions

Sl.No.	*Answer*	*Sl.No.*	*Answer*
1.	(d) all of these	**4.**	(d) Trade secret
2.	(d) indefinite period	**5.**	(c) Trade secret
3.	(c) Trade secret	**6.**	(c) Trade secret

D. True or False

1 to 11 all are true

E. Fill in the Blanks

Sl.No.	*Answer*	*Sl.No.*	*Answer*
1.	Primary	**7.**	Chemical
2.	Registration	**8.**	Commercial
3.	Company	**9.**	Reverse engineering
4.	Owner	**10.**	Owner
5.	Indefinitely	**11.**	Ideas
6.	Infringement		

F. Short or One Word Answer

Sl.No.	*Answer*	*Sl.No.*	*Answer*
1.	Trade secret	**4.**	Trademark
2.	Copyright	**5.**	Trade name
3.	Patent	**6.**	Domain name

CHAPTER 9
Related Rights

A. Definitions

Copyright: The exclusive rights granted to author/creator to reproduce, sell and distribute a work, prepare derivative works and display the work publicly.

Neighboring Right: Same as related rights.

Related Rights: Related rights is a term in copyright law, used in opposition to the term authors' rights.

Related right infringement: An unauthorized use of material protected by related rights.

B. Background Information

1. Related rights are rights which are similar to authors' right but which are not connected with the actual author of the work. Both authors' rights and related rights are copyrights in the sense of English or U.S. law.

2. **Matters Covered:** Related rights or neighbouring rights cover: performers rights, the rights of phonogram producers, publishers' rights in typographical arrangements, the rights of broadcasting organizations in relation to their broadcasts, film producers, database creators, photographers and designers.

3. **Period of Protection:** The period of protection varies with the matters protected by related rights and also with the countries. The period of protection as per Rome Convention for different matters is as shown in Table 9.1.

4. **Relation to authors' rights:** Related rights are independent of any authors' rights which might also exist in the work. Hence a CD recording of a song will be concurrently protected by four different copyright-type rights:

Table 9.1: Period of Protection for different Matters

Sl.No.	*Items/Matters*	*Period (Years)*	*Date of Commencement*
1.	Performers	15	From end of the year in which performance was made.
2.	Phonogram producers	20	From end of the year in which phonogram was first published. It is 50 years in European Union.
3.	Broadcasting organizations	20	From the end of the year in which the broadcast was first made.
4.	Film Producers	50	From the end of the year in which the film was first published or otherwise made available to the public, or for fifty years from the end of the year in which the master copy was made if the film is not released.
5.	Database creators	15	From the "completion" of the database.
6.	Photographers	25	From date of creation.
7.	Designers	25	After creation.
8.	Semiconductor designers	10-15	Ten years from the first commercial exploitation, or for fifteen years from the first creation.

(*i*) the authors' rights of the composer of the music;

(*ii*) the authors' rights of the lyricist;

(*iii*) the performers' rights of the singer and the musicians;

(*iv*) the producers' rights of the person or corporation which made the recording.

5. **Rights covered:** The owner of have the rights to prevent other to use their creations without their permission. The rights of different owners are briefly presented below:

(1) Performers: Under the Rome Convention, performers have the right to prevent:

the broadcast or communication to the public of their performance, unless this is made from a legally published recording of the performance;

the fixation (recording) of their performance;

the reproduction of a recording of their performance.

(2) Phonogram Producers: The producers of phonograms, that is the person who makes the recording rather than the person who performs, has the right to prevent the direct or indirect reproduction of the recording.

(3) Broadcasting Organizations: Rome Convention specifies that broadcasting organisations shall have the right to prohibit (or licence):

(*i*) the rebroadcasting of their broadcasts;

(*ii*) the fixation (recording) of their broadcasts;

(*iii*) the reproduction of fixations of their broadcasts;

(*iv*) the communication of their broadcasts to the public in places where an entrance fee is charged.

(4) Film Producers: Film producers have the right to prevent:

the direct or indirect reproduction of the film (either the master copy or copies thereof); the distribution of the film (or copies thereof) to the public for sale; the "making available to the public" of the film.

(5) **Database Creators:** Database creators have the right "to prevent extraction and/or re-utilization of the whole or of a substantial part, evaluated qualitatively and/or quantitatively, of the contents of that database."

Similarly photographers and designers have rights to prevent others to use their creations without permission.

6. **Authorization:** The owners of above mentioned property have the rights to authorise or to provide license to others to make use of their related rights on royalty basis. The performer has the right to licence:

(*i*) the distribution of recordings of their performance, for sale or other transfer of ownership;

(*ii*) the rental of recordings of their performances, unless there is a compulsory licence scheme in operation;

(*iii*) the "making available to the public" of their performances, in effect their publication on the internet.

The owner of the phonogram has the rights to licence:

the distribution of their phonograms, for sale or other transfer of ownership;

the rental of their phonograms, unless there is a compulsory licence scheme in operation;

the "making available to the public" of their phonograms, in effect their publication on the internet.

Once a phonogram has been published, the producer cannot prevent its broadcast: an equitable fee for the licence may be either agreed between phonogram producers and broadcasters or imposed by law.

7. **Registration:** Registration is not compulsory. However, registration is provides legal rights to prevent others to make unauthorised use of the inventions or creations of the owners. The registration is done in the Copyrights Office and prescribed fee is charged.

8. **Enforcement:** The related rights come into force after registration.Without registration, there is no protection of related rights.

9. **Infringement:** The unauthorized use of the protected matters is called infringement. The right holders can take legal action against thoe who are using their creations without permission. The penalties can be imposed against damages.

C. Multiple Choice Questions

1. **Related rights are similar to:**
   (a) Copyrights (b) Patents
   (c) Trademarks (d) Industrial designs

2. **Performers' rights come under:**
   (a) Copyrights (b) Patents
   (c) Trademarks (d) Related rights

3. **Performers' rights are protected for:**
   (a) 10 years (b) 15years
   (c) 20 years (d) 25 years

4. **Photographers' rights come under:**
   (a) Related rights (b) Patents
   (c) Trademarks (d) Copyrights

5. **Photographers' rights are protected for:**
   (a) 10 years (b) 15years
   (c) 20 years (d) 25 years

6. **Film producers' rights are protected for:**
   (a) 5 years (b) 10 years
   (c) 15 years (d) 20 years

7. **Designers' rights are protected for:**
   (a) 10 years (b) 15years
   (c) 20 years (d) 25 years

8. **Broadcasting rights are protected for:**
   (a) 20 years (b) 25years
   (c) 50 years (d) 70 years

9. **Designers' rights come under:**
   (a) Related rights (b) Patents
   (c) Trademarks (d) Copyrights

10. **Phonogram producers' rights come under:**
    (a) Related rights (b) Patents
    (c) Trademarks (d) Copyrights

11. Phonogram producers' rights are protected for:

(a) 15 years
(b) 20years
(c) 25 years
(d) 30 years

D. True or False

1. Related rights under primary intellectual property rights.
2. Related rights are also known as neighboring rights.
3. Broadcasting rights come under related rights.
4. Phonogram producers' rights come under related rights.
5. Performers rights come under related rights.
6. Film producers rights come under related rights.
7. Database creators' rights come under related rights.
8. Photographers' rights come under related rights.
9. Designers' rights come under related rights.
10. Related rights are independent of authors' rights.
11. Performers' rights can be protected for 15 years.
12. Database creators' rights can be protected for 15 years.
13. Phonogram producers' rights can be protected for 20 years.
14. Broadcasting' rights can be protected for 20 years.
15. Photographers' rights can be protected for 25 years.
16. Designers' rights can be protected for 25 years.
17. Film producers' rights can be protected for 50 years.
18. TRIPS agreement was signed in 1994.
19. Phonogram convention was held in Geneva in 1971.
20. Registration of related rights is done in copyright office.

E. Fill in the Blanks

1. Related rights under ________ intellectual property rights.
2. Related rights are also known as ________ rights.

3. Broadcasting rights come under ________ rights.
4. Phonogram producers' rights come under ________ rights.
5. Performers rights come under ________ rights.
6. Film producers rights come under ________ rights.
7. Database creators' rights come under ________ rights.
8. Photographers' rights come under ________ rights.
9. Designers' rights come under ________ rights.
10. Related rights are ________ of authors' rights.
11. Performers' rights can be protected for ________ years.
12. Database creators' rights can be protected for ________ years.
13. Phonogram producers' rights can be protected for ________ years.
14. Broadcasting' rights can be protected for ________ years.
15. Photographers' rights can be protected for ________ years.
16. Designers' rights can be protected for ________ years.
17. Film producers' rights can be protected for ________ years.
18. TRIPS agreement was signed in ________.
19. Phonogram convention was held in Geneva in ________.
20. Registration of related rights is done in ________ office.

F. Answer in Short or One Word

1. A term in copyright law, which is used in opposition to the term authors' rights
2. Neighboring rights
3. The exclusive rights granted to author/creator to reproduce, sell and distribute a work, prepare derivative works and display the work publicly.
4. An unauthorized use of material protected by related rights.

ANSWERS

C. Multiple Choice Questions

Sl.No.	*Answer*	*Sl.No.*	*Answer*
1.	(a) Copyrights	**7.**	(d) 25 years
2.	(d) Related rights	**8.**	(c) 50 years
3.	(b) 15years	**9.**	(a) Related rights
4.	(a) Related rights	**10.**	(a) Related rights
5.	(d) 25 years	**11.**	(b) 20years
6.	(d) 20 years		

D. True or False

1 to 20 all are true

E. Fill in the Blanks

Sl.No.	*Answer*	*Sl.No.*	*Answer*
1.	Primary	**11.**	15
2.	Neighboring	**12.**	15
3.	Related	**13.**	20
4.	Related	**14.**	20
5.	Related	**15.**	25
6.	Related	**16.**	25
7.	Related	**17.**	50
8.	Related	**18.**	1994
9.	Authors	**19.**	1971
10.	Independent	**20.**	Copyright

F. One Word Answer

Sl.No.	*Answer*	*Sl.No.*	*Answer*
1	Related rights	**3.**	Copyrights
2	Neighboring rights	**4.**	Related rights infringement

Chapter 10

Infringement or Piracy

A. Definitions

Bio-piracy: The illegal collection of indigenous plants by corporations who patent them for their own use.

Bio-prospecting: Bio-prospecting is the search for biological resourcesand related indigenous knowledge primarily for the purpose of commercial exploitation.

Copyright Infringement: The illegal use of a copyrighted work.

Client-Server Overuse Piracy: This type of piracy occurs when too many employees on a network are using a central copy of a program at the same time.

Hard Disk Loading: This occurs when a business who sells new computers loads illegal copies of software onto the hard disks to make the purchase of the machines more attractive.

End User Piracy: This occurs when a company employee reproduces copies of software without authorization.

Industrial Design Infringement: Unauthorized use of industrial design without the permission of owner.

Infringement: An unauthorized use of material protected by copyright, patent or trademark law.

Internet Piracy: This occurs when software is downloaded from the Internet.

Patent infringement: Unauthorized use of a patent.

Trademark infringement: Unauthorized use of a trademark.

Software Counterfeiting: This occurs when pirates deliberately and illegally duplicate and sell copyrighted material.

Software Piracy: Unauthorized duplication of computer software.

B. Background Information

1. The infringement of Intellectual Property Rights refers to unauthorized copying or use of a protected form of intellectual property.

2. It is an unauthorized use of material protected by copyright, patent or trademark law is called infringement. It amounts to exploitation of another individual or organization's intellectual property rights without lawful authority.

3. Piracy refers to an unauthorized duplication of goods protected by intellectual property law (*e.g.* copying software unlawfully). It refers to copying or distributing protected works or products without the permission of the true owner/author of an intellectual property.

4. Infringement may be of different types such (i) copyright infringement, (ii) patent infringement, (iii) trademark infringe,(iv) industrial design, infringement, (v) biopiracy, and (vi) software piracy. These are defined as follows:

5. The illegal use of a copyrighted work or violation of the exclusive rights granted by the copyright law is referred to as infringement of copyrights.

6. Making, using or selling an invention protected by a patent without the inventor's permission is called patent infringement. In other words, infringement of a utility patent involves the making, using, selling, offering to sell, or importing of a patented product or process without permission.

7. Violation of trademark rights through unauthorized use of a trademark is called trademark infringement. In other words, it refers to ihe improper use of another's trademark or a confusingly similar mark that creates the appearance of an affiliation with or actually being the other product or service.

8. Industrial Design Infringement refers to violation of industrial design rights through unauthorized use of a design or unauthorized use of industrial design without the permission of owner is called industrial design infringement.

9. The illegal collection of indigenous plants by corporations who patent them for their own use is called bio-piracy. It refers to:
   (*i*) Unauthorized use of biological resources such as plants, animals, micro-organisms and genes,
   (*ii*) Unauthorized use of traditional communities' knowledge on biological resources,
   (*iii*) Unequal share of benefits between a patent holder and the indigenous community whose resources and knowledge have been used; and

(*iv*) Patenting of biological resources with no respect to patentable criteria such as novelty, non-obviousness and usefulness.

Bio-piracy commonly operates through the application of Intellectual Property Rights to genetic resources and traditional knowledge. Bio-prospecting is the search for biological resources and related indigenous knowledge primarily for the purpose of commercial exploitation. The issue of bio-piracy is mostly raised by under-developed biodiversity rich countries such as India, Brazil and Malaysia among others.

10. Software piracy is the unauthorized duplication of computer software. In 2003, the estimated losses due to software piracy exceeded 29 billion dollars. Software piracy is of many forms such as (i) end user piracy, (ii) client-server overuse, (iii) internet piracy, (iv) hard disk loading, and (v) software counterfeiting. These forms are briefly discussed as follows.

(*i*) **End User Piracy** occurs when a company employee reproduces copies of software without authorization. End user piracy can take the following forms:

(1) Sing one licensed copy to install a program on multiple computers;

(2) Copying disks for installation and distribution;

(3) Taking advantage of upgrade offers without having a legal copy of the version to be upgraded;

(4) Acquiring academic or other restricted or non-retail software without a license for commercial use;

(5) Swapping disks in or outside the workplace.

(*ii*) **Client-Server Overuse** piracy occurs when too many employees on a network are using a central copy of a program at the same time. If you have a local-area network and install programs on the server for several people to use, you have to be sure your license entitles you to do so. If you have more users than allowed by the license, that's "overuse".

(*iii*) **Internet Piracy**

This occurs when software is downloaded from the Internet The same purchasing rules should apply to online software purchase as for those bought in traditional ways. Internet piracy can take the following forms:

(1) Pirate websites that make software available for free download or in exchange for ploaded programs;

(2) Internet auction sites that offer counterfeit, out-of-channel, infringing copyright software;

(3) Peer-to-Peer networks that enable unauthorized transfer of copyrighted programs.

(iv) **Hard-Disk Loading** occurs when a business who sells new computers loads illegal copies of software onto the hard disks to make the purchase of the machines more attractive. The same concerns and issues apply to Value Added Resellers (VAR) that sell or install new software onto computers in the workplace.

(v) **Software Counterfeiting** occurs when pirates deliberately and illegally duplicate and sell copyrighted material, often making it appear to their customers that they are purchasing an authentic product. When it comes to packaged software, like that sold by Adobe, it is common for pirates to market and sell counterfeit copies of the software CDs along with reproduced packaging, manuals, license agreements, labels, registration cards, and security features.

Illegally copied and sold software is not eligible for support, training, or upgrades. You may not be able to register it, so it may not work properly.

11. The penalties can be imposed for infringement of copyrights, patents, trademarks, industrial design and piracy of software.

(*i*) **Copyrights:** Penalties for copyright infringement can include:

An injunction against further infringement such as an order preventing, the infriger from further copying or distribution of the copyrighted work.

Impounding or destruction of infringing copies.

Charging Damages: Either actual damages and the infringer's profit or statutory damages are charged. Costs and attorney's fees.

A copyright owner can only sue for infringement on a work that was registered with copyright office. The action against has to be taken within three months of first publication.

(*ii*) **Patents:** A patent owner may recover money damages in the form of a reasonable royalty. If the infringement is found willful, three times money is recovered.

(*iii*) **Trademarks**: Same action as in case of patents can be taken against infringement of trademarks.

(*iv*) **Industrial Design**: The owner has the right to take legal action against the infringer and claim damages. However, the legal action must be taken within three years of the alleged offence.

(*v*) **Software:** There are significant financial and legal penalties for software piracy. Illegal distribution and use of software can result in prison terms of up to **five** years, and heavy fines [up to US$250,000 in USA]. In civil litigation against those who illegally distribute Adobe software, Adobe can obtain the higher of its lost profits, the infringer's profits, or statutory damages plus attorney fees. But the costs are not measured in money alone. When organizations are legally prosecuted for software piracy, they also suffer significant damage to their reputation.

C. Multiple Choice Questions

1. Infringement may occur with to:

(a) Copyrights (b) Patents

(c) Trademarks (d) All of these

2. Infringement may occur with to:

(a) Trade name (b) Industrial design

(c) Trademarks (d) All of these

3. Infringement may occur with to:

(a) Software (b) Utility model

(c) Trademarks (d) All of these

4. Software piracy can result in imprisonment up to:

(a) 1 year (b) 2 year

(c) 5 year (d) 7 year

5. Unauthorized use of Computer software refers to:

(a) Internet piracy (b) Software piracy

(c) Bio-piracy (d) All of these

6. Bio-piracy refers to unauthorized use of:

(a) Plants (b) Animals

(c) Micro-organisms (d) All of these

7. Bio-piracy gives importance to:

(a) Novelty (b) Non-obviousness

(c) Usefulness (d) All of these

8. Downloading of software from internet refers to:

(a) Internet piracy (b) Software piracy

(c) Bio-piracy (d) All of these

9. Penalties for infringement of copyright may include:

(a) Injunction (b) Destruction of copies

(c) Attorney's fee (d) All of these

10. Software piracy refers to:

(a) Internet piracy (b) End user piracy

(c) Client server overuse (d) All of these

D. True or False

1. Unauthorized use of protected IPR is called infringement.
2. Unauthorized use of copyrights is called copyrights infringement.
3. Unauthorized use of patented IPR is called patent infringement.
4. Unauthorized use of protected industrial design is called industrial design infringement.
5. Unauthorized use of protected trademark is called trademark infringement.
6. Unauthorized use of computer software is called software piracy.
7. Bio-piracy is related to PGR and Traditional Knowledge.
8. Down-loading of software from the internet is called internet piracy.
9. Unauthorized use of biological resources refers to bio-piracy.
10. Penalty is up to 5 year prison for illegal use of software.

E. Fill in the Blanks

1. Unauthorized use of protected IPR is called ________.
2. Unauthorized use of copyrights is called ________ infringement.
3. Unauthorized use of patented IPR is called ________ infringement.
4. Unauthorized use of protected industrial design is called ________ infringement.
5. Unauthorized use of protected trademark is called ________ infringement.
6. Unauthorized use of computer software is called ________.
7. Bio-piracy is related to ________ resources.
8. Down-loading of software from the internet is called ________ piracy.
9. Unauthorized use of biological resources refers to ________.
10. Penalty is up to ________ year prison for illegal use of software.

F. Answer in Short or One Word

1. An unauthorized use of material protected by copyright, patent or trademark law.
2. The illegal use of a copyrighted work
3. Making, using or selling an invention protected by a patent without the inventor's permission.
4. Violation of trademark rights through unauthorized use of a trademark.
5. Unauthorized use of industrial design without the permission of owner.
6. Unauthorized duplication of computer software.
7. Unauthorized reproduction of copies of software by employees of a company.
8. Downloading od a software from the Internet.
9. Using a central copy of a program at the same time by too many employees on a network.
10. Loading illegal copies of software onto the hard disks while selling a new computer to make the purchase of the machines more attractive.
11. Deliberate and illegal duplication and sale copyrighted material.
12. The illegal collection of indigenous plants by corporations who patent them for their own use.
13. Search for biological resources and related indigenous knowledge primarily for the purpose of commercial exploitation.

ANSWERS

C. Multiple Choice Questions

Sl.No.	Answer	Sl.No.	Answer
1.	(d) All of these	6.	(d) All of these
2.	(d) All of these	7.	(d) All of these
3.	(d) All of these	8.	(d) All of these
4.	(c) 5 year	9.	(d) All of these
5.	(d) All of these	10.	(d) All of these

D. True or False

1 to 10 all are true

E. Fill in the Blanks

Sl.No.	Answer	Sl.No.	Answer
1.	Infringement	6.	Software
2.	Copyright	7.	PGR
3.	Patent	8.	Internet
4.	Industrial design	9.	Bio-piracy
5.	Trademark	10.	5

F. Short or One Word Answer

Sl.No.	Answer	Sl.No.	Answer
1.	Infringement	8.	Internet piracy
2.	Copyright infringement	9.	Client server over use piracy
3.	Patent infringement	10.	Hard disk loading
4.	Trademark infringement	11.	Software counterfeiting
5.	Design infringement	12.	Bio-piracy
6.	Software piracy	13.	Bio-prospecting
7.	End user piracy		

Chapter 11

Domain Name

A. Definitions

Domain Names: A domain name is also called a Web address. A domain name is used as an address for your Web site and e-mail, just like www.janesbagels.com and orders@janesbagels.com.

Domain Pointer/Sub-domains: Your hosting account will allow you to point or assign several domains (*e.g.* janesbagels.com).

Domain Protect: Domain Protect adds an extra layer of protection to your domain name registration, guarding it against unauthorized domain name transfers.

Domain Name System (DNS): A local database that translates domain names to Internet Protocol (IP) addresses.

Extensions: Network Solutions offers a variety of domain name extensions. Protecting brand identity has become very important, so often customers will register multiple extensions and variations of their domain names.

Extranet: An extranet is a network that allows a company to share information with other businesses and customers.

Generic top-level domain (gTLD): It is a top level domain used by a particular class of organization. These are three or more letters long, and are named for the type of organization that they represent (for example,.com for commercial organizations).

Internet Protocol (IP) Numbers (IP addresses): A unique, numeric identifier used to specify hosts and networks.

Internet Service Provider (ISP): A person, organization, or company that provides access to the Internet.

Premium Domain Name: Top level domain names, or also known as premium domain names, are commonly known as a domain ending with the extension

".com". **Registration:** The process through which individuals and organizations obtain a domain name.

Register Domain: Establishing an Internet identity to have a domain name.

Registration Fee: The charge for registering, a domain name.

Registry: Keeping records of Internet addresses such as Internet Protocol (IP) numbers and domain names,

Renewal: The process of reinitiating a domain name's registration for a specified period of time.

Renewal Fee: The charge for renewing an existing domain name registration.

Renewal Notice: A notice sent to registrants 60 days before the renewal of registration.

Second Level Domain: In the Domain Name System (DNS), the next highest level of the hierarchy underneath the top level domains.

Sub-domain: Web addresses built upon an existing domain name.

Third Level Domain: In the Domain Name System (DNS, the next highest level of the hierarchy underneath the second level domains. It is that portion of the domain name that appears two segments to the left of the top level domain.

Top Level Domain (TLD): In the Domain Name System (DNS), the highest level of the hierarchy after the root.

Transfer: The process of changing the party whom is listed as the domain name registrant.

URL (Uniform Resource Locator): A pointer to a page on the WWW or other resource on the Internet.

B. Background Information

1. Domain refers to the unique name that identifies an Internet site. Domain names have two or more parts separated by dots. For example www.ask-edi.com or www.bcbsks.com The term **domain name** has multiple related meanings:

2. It identifies a web site on the internet (eg, company.com); also referred to as a URL.[Uniform Resource Locator]

3. It allows reference to Internet sites without knowing the true numerical address.

4. It refers to the official name of a computer connected to the Internet. (eg, "www.microsoft.com").

5. It is the unique name of a computer on the Internet that distinguishes it from other systems around the world.

6. They are sometimes colloquially (and incorrectly) referred to by marketers as "web addresses".

7. There are guiding lines which should be given due importance while selecting a domain name. Some tips for choosing a domain name are as follows.

   (*i*) **Smaller Size:** The domain should not be too long. The domain name can be registered up to 63 characters long, but it is not necessary to use all 63 characters. In the case of domain names the smaller the better. Of course short domain names are harder to find but if you can find one it has a number of advantages such as, it's easier to brand, easier to remember and it's harder to misspell.

   (*ii*) **Easy to remember:** Try to get a name that's easy to remember and sounds good. General words are best for this but since these are all gone at this stage you should look for non general names or combinations of general names which rhyme and sound good together. Take elephantdroppings.com for example that's an easy to remember name.

   (*iii*) **Relate to your business:** If you use keywords which are related to your business industry you will help your search engine rankings. What's even better is if you can get your actual business name or a name similar to your company name as your domain name, this way all your off-line branding will pay off online as Internet users should be able to guess your domain name from your business name.

   (*iv*) **Avoid similarities:** This means you should avoid choosing domain names that already have similar counterparts online, if you choose one of these people could end up on your competitors site rather than yours because they got confused over the two domain names.

   (*v*) **.com is best:** You should always aim for the.com version of a name, that's what most people will go to after doing a search or after typing something directly into their address bar, if you register.tv or.net or any of the other available extensions and your competition has the. com version then you've just lost a potential customer.

   (*vi*) **Use of Hyphen:** Use hyphens if your name is more than 3 words long (unless your name includes 3 short words). A name like carbuying. com doesn't really need a hyphen because it's readable the way it is. However a name like used-car-buying-scams.com looks better than usedcarbuyingscams.com since there are so many words in the name.

   (*vii*) **Use of multiple Extensions:** Protect your name by registering multiple extensions (.net, .org, *etc.*) This deters people from copying your name. For example, if you own garden-tips-for-us.com you can also register garden-tips-for-us.net and any other extension that's available.

(viii) Other Points: The following points should also be kept in mind in choosing a domain name.

(1) Choose an appropriate top-level domain or domains.

(2) Try for a variant of the name of your service or Web site.

(3) Avoid names that are too narrow in scope (*e.g.*, the name of a short-term project or of one event).

(4) Use Web-based wizards or other tools, such as DomainConquest, to look for unclaimed domain names that match key words.

(5) Avoid trademarks (even your own name, if it is someone else's trademark).

(6) Avoid made-up words and alternate spellings (*e.g.*, "pythagoris") (they can be annoying and hard to remember), unless they are well established (*e.g.*, the name of your company).

(7) Use at least one reasonably rare word (you have a better chance that the name is not already taken).

(8) Avoid top-domain-like names (this can create problems with name resolution).

(9) Don't use an embarrassing name or a name that may antagonize the wrong people.

(10) Don't use numeric digits at the beginning of the name (*e.g*, "525").

(11) Don't use non-alphanumeric characters. Most of them don't work over the Internet. Hyphens and underscores are technically OK, but may be difficult for users to remember correctly. A name should never begin with "xn—", which is reserved for encoding Internationalized Domain Names.

(12) Don't expect case to be preserved.

(13) For a large company, consider registering many variant forms and names of products, to avoid their use by competitors, critics, and purveyors of goods with which the company does not wish to be associated.

8. **Registration:** The process through which individuals and organizations obtain a domain name is called registration. Registration of a domain name enables the individual or organization to use that particular domain name for a specified period of time, provided certain conditions are met and payment for services is made. The important points about registration are listed below:

(*i*) Domain Names can be registered through many different companies known as registrars. The registrar you choose will ask you to provide various contact and technical information that makes up the registration. The registrar will then keep records of the contact information and submit the technical information to a central directory known as the "registry." This registry provides other computers on the

Internet the information necessary to send you e-mail or to find your web site. You will also be required to enter a registration contract with the registrar, which sets forth the terms under which your registration is accepted and will be maintained.

(*ii*) The application is to be filed in a prescribed format called registration forms. These forms are used to register new domain names, new contacts for domain names, and new hosts (name servers) as well as to update domain name, contact, and host records.

(*iii*) The prescribed fee is to be deposited. It can be deposited by credit card, demand draft or in cash.

(*iv*) The domain name is allotted and activated after the payment of fee is received in the registrar office.

(*v*) Generic top level domains are available for registration on a global basis.

9. **Registration Fee:** It refers to the charge for registering, a domain name. The registration fee covers the cost of processing the initial registration and maintaining the domain name record for two years. The registration fee varies from country to country and company to company. After the initial two-year period, the domain name is subject to renewal (re-registration) and the renewal (re-registration) fee will be due on an annual basis if the registrant wishes to renew the domain name's registration.

10. **Period of Registration:** A domain name is registered for a period of one year which has to be renewed every year. In other words, the domain name is renewed on yearly basis.

11. **Enforcement:** The domain name becomes operative immediately after registration. It remains active upto 30 days after the expiry of the registration date.

12. **Transfer:** It refers to the process of changing the party whom is listed as the domain name registrant. The party taking over the domain name is responsible for paying a new registration fee. Domain name transfers are processed using the Registrant Name Change Agreement (RNCA).

13. **Infringement:** The use of a domain name after the expiray of 30 days from the due date of registration amounts to infringement. It invites legal action against the unauthorized users.

14. A local database that translates domain names to Internet Protocol (IP) addresses is called domain name system. It is much easier for humans to remember and use names than IP numbers, which are what computers need to find each other on the Internet. Individual Internet computers around the globe maintain their specific portion of this database, and the data held in each portion of the database is made available to all

computers and users on the Internet. The DNS comprises computers, data files, software, and people working together.

15. Domain Names always have 2 or more parts separated by the dots. The leftmost piece is the most specific part while the right part is the TLD or Top Level Domain.Generally, there are three parts of a domain name such as (i) top level domain, (ii) second level domain, and (iii) third level domain. These are defined as follows.

    (*i*) **Top Level Domain (TLD):** The rightmost part of a domain name is called Top Level Domain TLD]. In other words, in a domain name, that portion of the domain name that appears in the rightmost position is known as top level domain. For example, the com in DomainMart.com. The main points related to TLD are given below:

    (1) It is the rightmost part of domain name.

    (2) Every domain name ends with top level domain [TLD].

    (3) It is also known as first level domain.

    (4) It generally consists of three generic characters.

    In the beginning there were only 5 Global top level domains [TLD's] that were recognized by everyone. These TLD's are given below:

    .com This TLD is intended for Commercial business around the world.

    .net This TLD is historically and is still used by Internet Service Providers.

    .org This TLD is intended to search the noncommercial community.

    .mil This TLD is restricted to use by the United States Military.

    .gov This TLD is restricted to use by United States Government Agencies.

    (*ii*) **Second Level Domain:** In the Domain Name System (DNS), the next highest level to the left position of the top level domains is referred to as second level domain. In other words, it is that portion of the domain name that appears immediately to the left of the top level domain. For example, the DomainMart in DomainMart.com.

    (*iii*) **Third Level Domain:** In the Domain Name System (DNS, the next highest level of the hierarchy underneath the second level domains is known as third level domain. It is that portion of the domain name that appears two segments to the left of the top level domain. Third level domains are not the portion of an email address that appears in front of the @ symbol - for example, the webmaster in support@DomainMart.com is not a third level domain.

    There can be **fourth and fifth level domains** and so on, with virtually no limitation. An example of a working domain with five levels is *www.sos.state.oh.us*. Each level is separated by a dot or period symbol between them.

Beside these, there are two more terms *viz.* generic top level domain, country code top level domain, subdomain and host nane. These are defined as follows.

Generic Top-Level Domain (gTLD): It is a top level domain of an Internet address used by a particular class of organization. These are three or more letters long, and are named for the type of organization that they represent The generic top level domains include .com, .net, .org, .mil, .gov, *etc.* In addition, seven new gTLDs were also selected by ICANN (the Internet Corporation for Assigned Names and Numbers) on November 16, 2000. These are: .aero (for the entire aviation community); .biz (for business purposes); .coop (for cooperatives); .info (unrestricted); .museum (for museums); .name (for personal names); .pro (for professionals).

Country Code Top Level Domain [ccTLD]: A ccTLD is a country code top-level domain, for example: .mx for Mexico. These ccTLDs are administered independently by nationally designated registration authorities. There are currently 243 ccTLDs reflected in the database of the Internet Assigned Numbers Authority (IANA). WIPO, which has a ccTLD Program, has launched a database portal, facilitating online searches for information related to country code top level domains.

(a) **Subdomain:** Generally known as a "Domain within a Domain", subdomains are web addresses built upon an existing domain name. In other words, a subdomain is a subdivision of a larger domain. For example, "mail.yahoo.com" is a subdomain of "yahoo.com."

(b) **Hostnames:** The hostname of an Internet Address is all of the left most pieces of a full internet address if there are more than 2 parts to the address. If there are only 2 parts of the address, then the hostname is equivalent to the domain name. Some examples are as follows:

Full Address	*Hostname*	*Domain Name*	*TLD*
www.keycomputer.com	www	keycomputer.com	com
keycomputer.com	keycomputer.com	keycomputer.com	com
www.google.net	www	google.net	net
my.www.keycomputer.com	my.www	keycomputer.com	com

(i) It is clear from the examples the hostname can be many different things.

(ii) If the entire address consists of only 2 parts, then the hostname is equivalent to the domain name.

(iii) If the entire address consists of more than 2 parts, then the hostname is equivalent to all the left most pieces before the right 2 parts (or the domain name).

In other words, a hostname refers to a domain name that has one or more associated Internet Protocol [IP] addresses. For example, the en.wikipedia. org and wikipedia.org domains are both hostnames, but the org domain is not.

16. A domain name extension is also called a top level domain name such as.com,.org,.net, and a lot more. It is the top most part of the web site address. The domain name extensions are mainly of three types, *viz.,* (i) International domain name extension, (ii) National domain name extension, and (iii) alternative domain name extension. These are explained as follows.

(1) **International Domain Extensions:** International Domain Extensions are also referred to as common domain extensions. Three domain name extensions, *viz.,*.com,.net, and.org are the most common on the international front. Moreover, these domain names are called as the "Big 3" because they accessible to anyone, without limitations. They are now widely recognized in all parts of the globe.

(2) **National Domain Extensions:** National Domain Extensions are also known as Country Code Domain Extensions. Now most countries now have a matching top level domain name, under which domain names are sold. The most common domain name extensions are based on two characters ISO country codes. Most popular country top level domains include the .fr for France, .jp for Japan, and.co.uk. These domain name extensions are generally under the control of the government and have high level of stability and durability. The Country Code Top Level Domains [ccTLD] are becoming more popular due to the reducing supply of the more common.com, .net and .org extensions.

(3) **Alternative Domain Extensions:** In the year 2000 the ICANN [Internet Corporation for Assigned Names and Numbers]approved a new set of seven top domains such as .aero, .biz, .coop, .info, .museum, .name, and .pro. The uses of some generic top level domains are explained as follows.

Table 11.1: Uses of some Generic Top Level Domains

Sl.No.	*Domain Extension*	*Meant for or Designated for*
1.	.com	For the commercial organizations.
2.	.net	For network services like Internet Service Providers.
3.	.org	For non-profit organizations.
4.	.aero	For entire aviation community.
5.	.biz	For business purposes.
6.	.coop	For cooperatives.
7.	.inf	For unrestricted use.
8.	.museum	For museums.
9.	.name	For personal names.
10.	.pro	For professionals.

They are generally called "alternative top level domain names". They are officially known, thus will work in all parts of the globe, in any browser or other software. These are also known as generic top level domains.

Recently, ten new domain name extensions have been included. These are are .asia, .cat, .jobs, .mail, .mobi, .post, .tel, .travel, and .xxx.

17. There are many types of domain extensions and each has its importance and value The importance of some domain extensions is given below:

(*i*) **".com"** for example is perceived in the public as more credible and usable than a ".tv" or ".mobi", it is more "official." In the academic world, the ".com", is preferred rather ".org" or ".edu".**com** These sites are generally businesses of some sort or informational portals. The most popular extension, this is the one people know about.

(*ii*) **.org, .edu** These sites are more credible and usually often are dedicated to a school or non-profit organization.

(*iii*) **.tv, .mobi, .us**, *etc.*, These extensions generally are different types of domain extensions associated with how you access them.**mobi**, for example, refers to accessing it through your mobile phone. They are not as hot as a.**com,** though still very popular and growing.

18. ICANN refers to the Internet Corporation for Assigned Names and Numbers. ICANN is responsible for managing and coordinating the Domain Name System (DNS) to ensure that every address is unique and that all users of the Internet can find all valid addresses. It does this by overseeing the distribution of unique IP addresses and domain names. It also ensures that each domain name maps to the correct IP address.

19. ICANN is also responsible for accrediting the domain name registrars. "Accredit" means to identify and set minimum standards for the performance of registration functions, to recognize persons or entities meeting those standards, and to enter into an accreditation agreement that sets forth the rules and procedures applicable to the provision of Registrar Services.

20. ICANN's role is very limited, and it is not responsible for many issues associated with the Internet, such as financial transactions, Internet content control, spam (unsolicited commercial email), Internet gambling, or data protection and privacy.

21. There are several advantages of domain names. Some advantages of domain names are as follows.

(*i*) **Naming Rights:** A domain name may only be registered once and for one buyer only. Since it is in a first come, first served basis, the first one to register the name shall have the right to use the name. It is a memorable website address. The owner has complete control over his website.

(*ii*) **Identity:** A domain name represents an identity which visitors always recognize. It is just like a brand name or a logo,

(*iii*) **Simplicity:** A domain name shortens the URL address of your blog and it is much easier to remember. It is easier for users to remember domain name than host-domain-path which is longer (such as *instruct.uwo.ca/~craven*).

(*iv*) **Adds Credibility:** A domain name projects a professional image of a owner and thus adds credibility to the owner. In other words, a domain name looks more professional and does have more perceived credibility from all your visitors.

(*v*) **More Advertisers:** Some advertisers prefer a domain name than a sub domain name. With the use of a domain name, there is an expansion of opportunities for your blog. More advertisers, more tasks.

(*vi*) **More impressive:** It is very impressive (looks less like a personal site)

(*vii*) **Transportable:** It is transportable to another hosting service.

(*viii*) Customer service is readily available.

(*ix*) Domain name makes the keywords clear to search engines because the words are separated.

22. Sometimes the name we want may already be taken; we may have to pay a lot (or sue) to obtain it. It requires proper maintenance of DNS records on a server.

C. Multiple Choice Questions

1. **A domain name ending with .com is called:**
   (a) Host name (b) Premium domain
   (c) Sub-domain (d) None of these

2. **A domain name ending with .biz is used for:**
   (a) Aviation (b) Business purpose
   (c) Internet services (d) none of these

3. **A domain name ending with .aero is used for:**
   (a) Aviation (b) Business purpose
   (c) Internet services (d) none of these

4. **A domain name ending with .com is used for:**
   (a) Commercial organizations (b) Business purpose
   (c) Cooperatives (d) none of these

5. **A domain name ending with .coop is used for:**
   (a) Aviation (b) Business purpose
   (c) Cooperatives (d) none of these

6. **A domain name ending with .net is used for:**
   - (a) Aviation
   - (b) Business purpose
   - (c) Internet services
   - (d) none of these
7. **A domain name ending with .museum is used for:**
   - (a) Aviation
   - (b) Business purpose
   - (c) Museum
   - (d) None of these
8. **A domain name ending with .gov is used for:**
   - (a) Aviation
   - (b) Business purpose
   - (c) Government organization
   - (d) none of these
9. **A domain name ending with .org is used for:**
   - (a) Aviation
   - (b) Nonprofit organizations
   - (c) Internet services
   - (d) none of these
10. **A domain name ending with .pro is used for:**
   - (a) Aviation
   - (b) Business purpose
   - (c) Professionals
   - (d) none of these

D. True or False

1. Domain name is also known as unique name.
2. Registration of domain name is compulsory.
3. Domain name becomes operative immediately after registration.
4. First renewal of domain name becomes due after 2 years.
5. Subsequent renewal is done on yearly basis.
6. Transfer of domain name is possible.
7. Every domain name ends with top level domain.
8. Top level domain is also known as first level domain.
9. Domain name ending with.com is known as premium domain.
10. Domain .pro is used by professionals.
11. Domain .gov is used by government offices.
12. Domain name identifies an internet site.

13 aero is used by aviation community.

14. biz is used for business purposes.
15. coop is used by cooperatives.

E. Fill in the Blanks

1. Domain name is also known as ________.
2. Registration of domain name is ________.
3. Domain name becomes operative immediately after ________.
4. First renewal of domain name becomes due after ________ years.
5. Subsequent renewal is done on ________ basis.
6. Transfer of domain name is ________.
7. Every domain name ends with ________ domain.
8. Top level domain is also known as ________ domain.
9. Domain name ending with.com is known as ________ domain.
10. Domain .pro is used by ________.
11. Domain .gov is used by ________.
12. Domain name identifies an ________.
13. aero is used by ________ community.
14. biz is used for ________ purposes.
15. coop is used by ________.

F. Answer in Short or One Word

1. A name which is used as an address for your Web site and e-mail.
2. An extra layer of protection added to your domain name registration, guarding it against unauthorized domain name transfers.
3. A network that allows a company to share information with other businesses and customers.
4. Establishing an Internet identity to have a domain name.
5. A local database that translates domain names to Internet Protocol (IP) addresses
6. A top level domain used by a particular class of organization.

7. A unique, numeric identifier used to specify hosts and networks.
8. A person, organization, or company that provides access to the Internet.
9. A domain name ending with the extension ".com".
10. The process through which individuals and organizations obtain a domain name.
11. A registry is responsible for delegating Internet addresses such as Internet Protocol (IP) numbers and domain names, and keeping a record of those addresses and the information associated with their delegation.
12. The process of reinitiating a domain name's registration for a specified period of time.
13. The charges paid for renewing an existing domain name registration.
14. A notice sent to registrants 60 days before the expiry of their registration date.
15. In a domain name, that portion of the domain name that appears immediately to the left of the top level domain.
16. Web addresses which are built upon an existing domain name.
17. That portion of the domain name that appears two segments to the left of the top level domain.
18. In a domain name, that portion of the domain name that appears furtherest to the right.
19. The process of changing the party whom is listed as the domain name registrant.
20. A pointer to a page on the WWW or other resource on the Internet.

ANSWERS

C. Multiple Choice Questions

Sl.No.	Answer	Sl.No.	Answer
1.	(b) Premium domain	6.	(c) Internet services
2.	(b) Business purpose	7.	(c) Museum
3.	(a) Aviation	8.	(c) Government organization
4.	(a) Commercial organizations	9.	(b) Nonprofit organizations
5.	(c) Cooperatives	10.	(c) Professionals

D. True or False

1 to 15 all are true

E. Fill in the Blanks

Sl.No.	Answer	Sl.No.	Answer
1.	Unique name	9.	Premium
2.	Compulsory	10.	Professional
3.	Registration	11.	Government
4.	2	12.	Internet site
5.	Yearly	13.	Aviation
6.	Possible	14.	Business
7.	Top level	15.	Co-operatives
8.	First level		

F. Short or One Word Answer

Sl.No.	Answer	Sl.No.	Answer
1.	Domain name	11.	Registry
2.	Domain protect	12.	Renewal
3.	Extranet	13.	Renewal Fee
4.	Register domain	14.	Renewal Notice
5.	Domain name system	15.	Second level domain
6.	Generic top level domain	16.	Sub-domain
7.	Internet protocol	17.	Third level domain
8.	Internet service provider	18.	Top level domain
9.	Premium domain name	19.	Transfer
10.	Registration	20.	URL

Chapter 12

Trade Name

A. Definitions

Juridical Person: A partnership, corporation, cooperative, and legal labor organization, *etc.* that are registered in the particular concerned agencies.

Natural Person: Sole proprietorship/individual.

Trademarks: A symbol used to identify and distinguish companies, partnerships, and businesses, as opposed to marks used to identify and distinguish goods or services.

Trade Name: The name under which a company conducts business, or by which its business, product/service are identified. It may or may not be registered as a trademark.

Business Name: Same as trade name.

Brand Name: Same as trade name.

Corporate Name: Same as trade name.

Fictitious Name: An assumed name.

Trading name: Same as trade name.

B. Background Information

1. A trade name is the name under which a product is commercially known. The commercial name by which a chemical is known is called trade name. One chemical may have a variety of trade names depending on the manufacturers or distributors involved.

2. It is also known as business name, trading name, assumed name, brand name and corporate name.

3. It may or may not be registered.

4. It distinguishes a particular business from others.

5. It can be exclusive or non-exclusive.

6. It is generally used on letter heads and bank accounts.

7. It is the name given to a particular substance by each company that manufactures it.

8. It is the name under which a company conducts its business.

9. It is s the business name of the person or organization making and/or selling a particular product. "General Motors, Inc" is the "trade name" of the company

10. A business name must have distinctive, descriptive and legal elements and each element is considered as part of the whole name. The name should not be too general.

11. **Distinctive:** This is the element which is the unique identifier of the business. This element is the main identifier of your business name. This is the part of the name that sets it apart from other businesses of the same type.

12. **Descriptive:** This is the element which describes the main aspect of the business.

13. **Legal:** This is the element when the name is used by a corporation. The legal element indicates the status of the organization as a corporate body.

14. A business name should not suggest or imply a connection with any government or government department.

15. A business name must not suggest a connection with the Royal Family. The name of a society must be such that it will not be confused with existing names of societies, corporations, partnerships or trade names.

16. There are three types of business ownership or business structures which are most commonly used. These are Sole Proprietorship, Partnership, and Company or Corporation.

17. **Sole Proprietorship** is the simplest form of business in which one person performs all the functions that are required for the successful operation of a business. The proprietor secures the capital, hires any employees, establishes and operates the business, assumes all the risks, accepts all profits or losses and pays all the taxes. The proprietor is said to be self-employed.

18. Liability arising out of the business activity is borne entirely by the proprietor. All assets, whether business or personal, can be legally taken and used to discharge the liability.

19. A sole proprietorship can be operated under the individual's name without the need for registration.

20. The business ends with the death or incapacity of owner. Therefore, this type of business cannot be inherited. However, the individual assets can be used to carry on a business.

21. A partner-ship is a voluntary contract between two or more competent persons to place their money, efforts, labour and skill in business with the understanding that there shall be a proportional sharing of profits and losses between them. A partnership is governed by the *Partnership Act.*

22. A corporation or company is a separate legal entity having an existence separate and apart from the business person that created it. A company can be formed under the Company Act or the Corporations Act. The provincial corporations are most common. Its continuance depends only on the filing of annual returns and annual meetings of shareholders and directors. The Company Act allows a company to be owned by just one person. If more than one person is involved in the business, a Unanimous Shareholder's Agreement is recommended to establish the operational rules between the shareholders including dispute resolution and share evaluation. It is wise to consult with a lawyer to ensure that the company is set up in a manner that meets your business' present and future needs.

23. A business name registration is valid for three years from the date of registration. At that time, registration will expire and renewal of registration must be completed by an officer of the corporation or the business name will be removed.

24. When a corporation discontinues the use of a trade name, the name can be removed from the business names database. The name can be removed on written request only by the registered owners of the business name.

25. It may or may not be registered. However, registration is necessary to get legal rights. Trade names are registered by the state in which the person or company is based. For each State, it has to be registered separately. The application is filed in the prescribed form and prescribed processing fee is to be deposited.

26. Matters Covered: Trade names cover all types of business, *viz.* sole proprietership, partnership; corporate business, *etc.*

27. Duration of Protection: Initially the trade name is registered for a term of three years which is renewable for a further term of three years each.

28. **Rights Covered:** The sole owner of the trade name has all the rights related to his business, partners have equal rights and in a corporation, board of Directors have the rights.

29. **Requirements:** The basic requirements for registering a trade name are that that the name should be new and distinctive. It should not confused with the trade name of any other person or corporation.

30. **Authorization/Transfer:** The owner of the trade name has right to transfer his business to any other person who is interested in the business. For transfer the request is submitted in the prescribed form and transfer fee is to be paid.

31. **Enforcement:** The trade name comes into force immediately after registration and it expires after death of the owner or after its term if not renewed timely.

32. **Infringement:** Unauthorised use of a trade name by another without permission amounts to infringement. The trade name owner has rights to prevent such use. He can take legal action against infringer.

C. Multiple Choice Questions

1. **Trade name of a company identifies its:**
   (a) Business (b) Product
   (c) Service (d) All of these

2. **Trade name is also known as:**
   (a) Business (b) Trading name
   (c) Brand name (d) All of these

3. **A business structure refers to:**
   (a) Sole trader (b) Partnership
   (c) Company (d) All of these

4. **Initially the trade name is registered for:**
   (a) 1 year (b) 2 year
   (c) 3 year (d) 5 year

5. **Trade name is renewable after:**
   (a) 2 year (b) 3 year
   (c) 4 year (d) 6 year

6. **Board of Directors has the right in:**
   (a) Sole Proprietorship (b) Partnership
   (c) Company (d) All of these

7. **Trade name covers:**
   (a) Sole Proprietorship (b) Partnership
   (c) Company (d) All of these

8. **For registration a trade name must have:**
   (a) Distinctiveness (b) Descriptiveness
   (c) Legal elements (d) All of these

9. **In a sole business, the proprietor:**
   (a) Secures capital (b) Establishes business
   (c) Operates business (d) All of these

10. **In partnership several persons put together:**
   (a) Money (b) Labor
   (c) Skill (d) All of these

D. True or False

1. Trade name is related to a company conducting a business.
2 Trade name rights come under primary intellectual property rights,
3. The trade name comes into force immediately after registration.
4. Trade name expires after death of the owner.
5. Trade name also expires if it is not renewed timely.
6. The trade name owner has right to transfer his business to any other person.
7. Initially the trade name is registered for a term of three years.
8. Infringement of trade name is possible.
9. Trade name distinguishes a particular business from others.
10. A trade name is renewable after 3 years.
11. Registration of trade name is not compulsory.
12. A business controlled by one person is called sole proprietorship.
13. A business controlled by two or more persons is called partnership.
14. A business controlled by board of directors is called company or corporation.

E. Fill in the Blanks

1. A company conducts its business in a ________.
2. Trade name rights come under ________ intellectual property rights,
3. The trade name comes into force immediately after ________.
4. Trade name ________ after death of the owner.
5. Trade name also expires if it is not ________ timely.
6. The trade name owner has right to transfer his ________ to any other person.
7. Initially the trade name is registered for a term of ________ years.
8. Infringement of trade name is ________.
9. Trade name distinguishes a particular ________ from others.
10. A trade name is renewable after ________ years.
11. Registration of trade name is not ________.
12. A business controlled by one person is called ________.
13. A business controlled by two or more persons is called ________.
14. A business controlled by board of directors is called ________.

F. Answer in Short or One Word

1. A partnership, corporation, cooperative, and legal labor organization, *etc.* that are registered in the particular concerned agencies.
2. Sole proprietorship/individual
3. A symbol used to identify and distinguish companies, partnerships, and businesses, as opposed to marks used to identify and distinguish goods or services.
4. The name under which a company conducts business, or by which its business, product/service are identified. It may or may not be registered as a trademark.

ANSWERS

C. Multiple Choice Questions

Sl.No.	*Answer*	*Sl.No.*	*Answer*
1.	(d) All of these	**6.**	(d) All of these
2.	(d) All of these	**7.**	(d) All of these
3.	(d) All of these	**8.**	(d) All of these
4.	(c) 3 year	**9.**	(d) All of these
5.	(b) 3 year	**10.**	(d) All of these

D. True or False

1 to 14 all are true

E. Fill in the Blanks

Sl.No.	*Answer*	*Sl.No.*	*Answer*
1.	Trade name	**8.**	Possible
2.	Primary	**9.**	Business
3.	Registration	**10.**	3
4.	Expires	**11.**	Compulsory
5.	Renewed	**12.**	Sole proprietorship
6.	Business	**13.**	Partnership
7.	3	**14.**	Company/Corporation

F. Short or One Word Answer

Sl.No.	*Answer*	*Sl.No.*	*Answer*
1.	Juridical person	**3.**	Trademark
2.	Natural person	**4.**	Trade name

Section II

Sui-Generis Rights

Chapter 13

Sui-Generis Intellectual Property Rights

A. Definitions

Database Rights: A 'database right' is an intellectual property right given to a computer database.

Farmers' Rights: The legal rights provided to farmers to save, use, sow, replant, exchange, share or sell his farm produce including seed of a variety protected under Plant Variety Protection Act.

Mask Work: A mask work is a two or three-dimensional layout of an integrated circuit

Moral Rights: Moral rights are a special extension to copyright that gives the originator of a copyrighted work rights over its use.

Plant Breeders' Rights: Plant breeders' rights, also known as plant variety rights (PVR), are intellectual property rights granted to the breeder of a new variety of plant.

***Sui-generis* Rights:** *Sui-generis* refers to things of their own kind or things with unique characteristics

Supplementary Protection Certificate: This term is used in European Countries. This is an extension of the term given to pharmaceutical or plant protection patent.

Traditional Knowledge: The Traditional Knowledge [TK] refers to the knowledge, innovations and practices of indigenous people and local communities.

B. Background Information

1. *Sui-generis* is a Latin expression, literally meaning of its own kind/genus or unique in its characteristics. In law, it is term of art used to identify

a legal classification that exists independently of other categorizations because of its uniqueness or due to the specific creation of an entitlement or obligation.

2. In Intellectual Property there are rights which are known as being *sui-generic* to owners of a small class of works. Such Intellectual Property Rights include (i) database Rights, (ii) mask work, (iii) plant breeders' rights, (iv) farmers' rights, (v) moral rights, (vi) supplementary protection certificate, and (vii) indigenous intellectual property,

3. A 'database right' is an intellectual property right given to a computer database.

Table 13.1: Intellectual Property Law: List of Sui Generic Rights

Sl.No.	*Items of Sui Generic Right*	*Definition or Brief Description*
1.	Database rights	A 'database right' is an intellectual property right given to a computer database.
2.	Mask Work	A mask work is a two or three-dimensional layout of an integrated circuit (IC or "chip").
3.	Plant Breeders' Rights	Rights granted to the breeder of a new variety of plant.
4	Farmers' Rights	The legal rights provided to farmers to save, use, sow, replant, exchange, share or sell his farm produce.
5	Moral Rights	Moral rights are a special extension to copyright that gives the originator of a copyrighted work rights over its use.
6.	Supplementary protection certificate	A supplementary protection certificate (SPC) is a *sui-generis*, patent-like, intellectual property right that is available for medicinal and plant protection products.
7.	Indigenous Intellectual Property	The knowledge, innovations and practices of indigenous people and local communities.

4. A mask work is a two or three-dimensional layout of an integrated circuit (IC or "chip"), *i.e.* the arrangement on a chip of semiconductor devices such as transistors and passive electronic components such as resistors and interconnections.

5. Plant breeders' rights, also known as plant variety rights (PVR), are intellectual property rights granted to the breeder of a new variety of plant. Plant Breeders Rights are granted to novel plant varieties that are distinctive, uniform, and stable (eg, cultivars bred true-to-type for desired traits).

6. The legal protection of a new plant variety is granted to the breeder or his successor. The effect of PBR is that prior authorization is required before the material can be used for commercial purposes.

7. Farmers' rights refer to the rights arising from the past, present and future contributions of farmers in conserving, improving and making available

plant or animal genetic resources, particularly those in the centres of origin/diversity.

8. In other words, the legal rights provided to farmers to save, use, sow, replant, exchange, share or sell his farm produce including seed of a variety protected under Plant Variety Protection Act refer to Farmers Rights. The purpose of these rights is to "ensure full benefits to farmers and support the continuation of their contributions.

9. The FAO Conference held in Rome from 11-29 November, 1989 endorsed the concept of Farmers Rights with a view to:
   (*i*) Ensuring global recognition of the need for conservation and the availability of sufficient funds for these purposes;
   (*ii*) Assisting farmers and farming communities throughout the world, especially those in areas of original diversity of plant genetic resources, in the protection and conservation of their PGR and of the natural biosphere; and,
   (*iii*) Allowing the full participation of farmers, their communities and countries in the benefits derived, at present and in the future, from the improved use of PGR.

10. Moral rights are a special extension to copyright that gives the originator of a copyrighted work rights over its use. The originator can specify moral conditions on the use or exploitation of their creation, even when the rights are licensed or sold to others'.

11. A supplementary protection certificate (SPC) is a *sui-generis*, patent-like, intellectual property right. This type of right is available for medicinal products, such as drugs, and plant protection products, such as insecticides, and herbicides.

12. The Indigenous Knowledge also known as Traditional Knowledge [TK] refers to the knowledge, innovations and practices of indigenous people and local communities.

13. TK is the: knowledge gained through tradition from ancestors about plants and animals, medicines, agricultural practices and other matters.

C. Multiple Choice Questions

1. Database rights last for:

(a) 5 years (b) 10 years
(c) 15 years (d) 20 years

2. Registered mask work rights last for:

(a) 5 years (b) 10 years
(c) 14 years (d) 15 years

3. **Unregistered mask work rights last for:**
   (a) 2 years (b) 5 years
   (c) 7 years (d) 10 years

4. **Moral rights are provided in addition to:**
   (a) Patent (b) Trademark
   (c) Copyrights (d) None of these

5. **Which of the following cannot be transferred?**
   (a) Patent (b) Trademark
   (c) Copyrights (d) Moral rights

6. **Which of the following comes under *sui-generis* rights?**
   (a) Patent (b) Trademark
   (c) Moral rights (d) Industrial designs

7. ***Sui-generis* rights include:**
   (a) Database rights (b) Mask work rights
   (c) Moral rights (d) All of these

8. **Which of the following does not come under *sui-generis* rights?**
   (a) Copyrights (b) Mask work
   (c) Moral rights (d) Database rights

D. True or False

1. A database right is granted to computer database.
2. Database rights last for 15 years.
3. Mask work is related to IC chips.
4. Unregistered mask work lasts for 2 years.
5. Registered mask work lasts for 10 years.
6. Moral rights are provided in addition to copyrights.
7. Moral rights cannot be transferred.
8. SPC is given to pharmaceutical or plant protection patent.
9. Maximum life of SPC is 5 years.
10. SPC can be issued only after 5 years of original patent.
11. SPC is an extension of a patent.

12. Plant breeders' rights come under *sui-generis* rights.
13. Plant breeders' rights are related to new plant variety.
14. Traditional knowledge is also known as Indigenous knowledge.

E. Fill in the Blanks

1. A database right is granted to ________ database.
2. Database rights last for ________ years.
3. ________ are related to IC chips.
4. Unregistered mask work lasts for ________ years.
5. Registered mask work lasts for ________ years.
6. Moral rights are provided in addition to ________.
7. Moral rights cannot be ________.
8. ________ is given to pharmaceutical or plant protection patent.
9. Maximum life of SPC is ________ years.
10. SPC can be issued only after ________ years of original patent.
11. SPC is an extension of a ________.
12. Plant breeders' rights come under ________ rights.
13. Plant breeders' rights are related to ________ plant variety.
14. Traditional knowledge is also known as ________ knowledge.

F. Answer in Short or One Word

1. Things of their own kind or things with unique characteristics
2. An intellectual property right given to a computer database.
3. A two or three-dimensional layout of an integrated circuit
4. Special extension to copyright that gives the originator of a copyrighted work rights over its use.
5. An extension of the term given to pharmaceutical or plant protection patent.
6. The knowledge, innovations and practices of indigenous people and local communities.

7. Intellectual property rights granted to the breeder of a new variety of plant.

8. The legal rights provided to farmers to save, use, sow, replant, exchange, share or sell his farm produce including seed of a variety protected under Plant Variety Protection Act.

ANSWERS

C. Multiple Choice Questions

Sl.No.	Answer
1.	(c) 15 years
2.	(b) 10 years
3.	(a) 2 years
4.	(c) Copyrights
5.	(d) Moral rights
6.	(c) Moral rights
7.	(d) All of these
8.	(a) Copyrights

D. True or False

1 to 14 all are true

E. Fill in the Blanks

Sl.No.	Answer
1.	Computer
2.	15
3.	Mask work
4.	2
5.	10
6.	Copyrights
7.	Transferred
8.	SPC
9.	5
10.	5
11.	5 years
12.	*Sui-generis*
13.	New
14.	Indigenous

F. Short or One Word Answer

Sl.No.	Answer
1.	*Sui-generis* rights
2.	Database rights
3.	Mask Rights
4.	Moral Rights
5.	Supplementary protection certificate
6.	Traditional knowledge
7.	Plant breeders' rights
8.	Farmers' rights

Chapter 14

Database Rights

A. Definitions

Data: This term is legally defined according to the institution. It generally refers to recorded information regardless of form. Most institutions hold title to data while researchers have rights to access the data.

Database: A collection of data organized for rapid search and retrieval by a computer.

Database Rights: An intellectual property right given to a computer database.

Data Management: The process of handling the database by central database management system.

Data Definition Language (DDL): The language used to define objects in a database: CREATE TABLE, CREATE INDEX, and so on.

Data Manipulation Language (DML): The language used to manipulate objects in a relational database:

Data model: A data model is **not** the model (conceptual/logical/physical) of your future or existing database: It is a generic model which you base your analysis and design upon.

Database model: A Database model is the logical conversion of an Entity Relationship model. It includes making tables, indexes and other mechanisms in the database, necessary to construct the database.

Database Normalization: Database Normalization is the process of analyzing your database model to ensure that information is stored only in one place in the database, and that there are no derivatives of the information within the database.

Data-type: Identifies the kind of information that an Attribute/column in an entity/table on a specific database platform represents.

Extraction: The permanent or temporary transfer of all or a substantial part of the contents of a database to another medium by any means or in any form.

Occurrence: Each row in a database table. If you have 100 customers in your CUSTOMERS table, the table has 100 occurrences.

Re-utilization: Any form of making available to the public all or a substantial part of the contents of a database by distributing of copies, by renting, by on-line or other forms of transmission.

RDBMS: Short for **R**elational **D**atabase **M**anagement System. A type of database management system (DBMS) that stores data in the form of related tables.

B. Background Information

1. The database refers to an organized collection of information, data, or citations stored in electronic format that can be searched for specific information or records by techniques specific to each database.

2 A 'database right' is an intellectual property right given to a computer database. A database is defined as a collection of independent works, data or other materials which: (a) are arranged in a systematic or methodical way; and (b) are individually accessible by electronic or other means.

3. **Registration:** In many respects, a database right is similar to copyright. It is created automatically. In other words, this is an automatic right which protects the investment (*i.e.* time, money and energy) that goes into obtaining, verifying or presenting the contents of a database. There is no need of registration of database. But it is advisable to register the databse, because registration provides legal rights against unauthorised use of database by other persons. It is a right gainst use an dnot a monopoly.

4. **Requirement:** The owner of a protected database is a person who has taken the risk of investing in the creation of database. For the database rights to apply, the database must be author's own intellectual creation and the result of substantial investment (financial, technical or otherwise). A database will qualify for protection if it meets the following requirements:

5. **Author's own Creation:** The database is a "collection of individual works, data or other materials arranged in a systematic or methodical way", and

   A substantial investment. There should be a proof that substantial investment has been made in creation and maintenance of database.

6. **Matters Covered:** Any work that shows originality in its selection, coordination and arrangement can be protected under database rights The following items are covered by database rights.
   1. Telephone directories or telephone book,
   2. Railway tome tables,
   3. Electronic photo book,

4. Encyclopedia with different entries,

5. Microfilm collections, *etc.*

The following items are not covered by database rights

(*i*) Any computer software used to access or maintain the database.

(*ii*) A movie, a recording, an audiovisual, cinematographic, literaryor musical work,

7. Duration of Protection: The term for protection of database rights is 15 years from the end of the year in which the database was completed or first made avilable to the public.The term is extended automatically, if a substantial change [additions, deletions or alterations] is made to the datbase during the term of protection The changes include substantial alterations, deletions and additions to the database in such a way so as to reflect substantial new investment.

8. Rights Provided: A database right prevents copying of substantial parts of a database (including frequent extraction of insubstantial parts). However, unlike copyright the cover is not over the form of expression of information but of the information itself. The database owner is entitled to prevent others from:

(*i*) extraction and or re-utilization of the whole or of a substantial part, evaluated qualitatively and/or quantitatively, of the contents of that database'.

(*ii*) repeated and systematic extraction and/or re-utilization of insubstantial parts of the contents of the database implying acts which conflict with a normal exploitation of that database or which unreasonably prejudice the legitimate interests of the maker of the database'.

9. Authorization: The owner of the database has the rights to authorize other person to use his database on the basis of royalty payment. In other words, the holder of database has right to sale or license his database.

10. Enforcement: The database rights come into force from the end of the year in which the the making of database was completed or the database was made available to the public.

11. Infringement: The rights holders may take action if their database is copied or used without permission. Database right can be infringed by:

(*i*) Copying the contents of a database and rearranging them.

(*ii*) Making the contents of the database available to the public without permission.

(*iii*) Unauthorized repeated and systematic extraction of 'insubstantial parts' of database.

(iv) Permanent or temporary transfer of the database contents to another medium by any means, or in any form.

12. As with copyright, if a database right is infringed a range of civil remedies are available to the database right holder including injunctions, an account of profits, damages and delivering up of goods.

13. The extraction or reutilization of substantial part of a published data is permitted without permission of the database creator under following circumstances:

(i) If the data is used for teaching or research and not for any commercial purpose.

(ii) When the identity of the data creator is not ascertainable or

(iii) When the term of database right is expired.

14. Database right is in many ways very similar to copyright so that, for example, there is no registration for database right – it is an automatic right and commences as soon as the material that can be protected exists in a recorded form.

15. Ownership of database rights is determined in the same way as copyright ownership. As with copyright, database rights can be sold or licensed.

16. Database right concerns control over the extraction and re-utilization of the contents of the database.

17. Since the whole data resides in one central database, the various programs in the application can access data in different data files. Hence data present in one file need not be duplicated in another. This reduces data redundancy. However, this does not mean all redundancy can be eliminated.

18. Reduced data redundancy leads to better data consistency.

19. Since related data is stored in one single database, enforcing data integrity is much easier. Moreover, the functions in the DBMS can be used to enforce the integrity rules with minimum programming in the application programs.

20. Related data can be shared across programs since the data is stored in a centralized manner. Even new applications can be developed to operate against the same data.

21. Maintenance is less and easy, again, due to the centralized nature of the system.

22. Database systems are complex, difficult, and time-consuming to design Substantial hardware and software start-up costs.

C. Multiple Choice Questions

1. **Database rights last for:**
   (a) 5 years (b) 10 years
   (c) 15 years (d) 20 years

2. **Database rights comes under:**
   (a) Primary rights (b) Ordinary rights
   (c) *Sui-generis* rights (d) All of these

3. **Which of the following is *sui-generis* right?**
   (a) Copyright (b) Patent
   (c) Database (d) Trademark

4. **Telephone directory comes under:**
   (a) Copyright (b) Database
   (c) Trademark (d) Trade secret

5. **Railway time table comes under:**
   (a) Patent (b) Trade name
   (c) Database (d) Trade secret

6. **Encyclopedia comes under:**
   (a) Copyright (b) Database
   (c) Trademark (d) Trade secret

7. **Electronic photo book comes under:**
   (a) Copyright (b) Database
   (c) Trademark (d) Trade secret

8. **Creation of database involves lot of:**
   (a) Finance (b) Technical skill
   (c) Energy (d) All of these

9. **Creation of database involves lot of:**
   (a) Money (b) Ideas
   (c) Time (d) All of these

10. **Central database systems are:**
    (a) Complex (b) Difficult
    (c) Costly (d) All of these

D. True or False

1. A database right is granted to computer database.
2. Database rights last for 15 years.
3. Database rights come under *sui-generis* intellectual property rights.
4. Registration of database is not compulsory.
5. Telephone directory comes under database.
6. Railway Time Table comes under database.
7. Encyclopedia comes under database.
8. Electronic photo book comes under database.
9. In UK database rights were first adopted in 1998.
10. A database right arises automatically.
11. The holder of database has right to sale or license his database.
12. Database is liable to infringement.

E. Fill in the Blanks

1. A database right is granted to ________ database.
2. Database rights last for ________ years.
3. Database rights come under ________ intellectual property rights.
4. Registration of database is not ________.
5. Telephone directory comes under ________.
6. Railway Time Table comes under ________.
7. Encyclopedia comes under ________.
8. Electronic photo book comes under ________.
9. In UK database rights were first adopted in ________.
10. A database right arises ________.
11. The holder of database has right to ________ his database.
12. Database is liable to ________.

F. Answer in Short or One Word

1. A recorded information regardless of form.
2. A collection of data organized for rapid search and retrieval by a computer.
3. An intellectual property right given to a computer database.
4. The process of handling the database by central database management system.
5. The language used to define objects in a database:
6. The language used to manipulate objects in a relational database:
7. A generic model which you base your analysis and design upon.
8. The process of analyzing your database model to ensure that information is stored only in one place in the database, and that there are no derivatives of the information within the database.
9. The kind of information that an Attribute/column in an entity/table on a specific database platform represents.
10. The permanent or temporary transfer of all or a substantial part of the contents of a database to another medium by any means or in any form.
11. Any form of making available to the public all or a substantial part of the contents of a database by distributing of copies, by renting, by on-line or other forms of transmission.
12. A type of database management system (DBMS) that stores data in the form of related tables.
13. The documentation stored in the database repository, and which holds information about your database objects.
14. Each row in a database table.
15. A general interface for communication with different vendor-specific relational database systems.

ANSWERS

C. Multiple Choice Questions

Sl.No.	Answer	Sl.No.	Answer
1.	(c) 15 years	6.	(b) Database
2.	(c) *Sui-generis* rights	7.	(b) Database
3.	(c) Database	8.	(d) All of these
4.	(b) Database	9.	(d) All of these
5.	(c) Database	10.	(d) All of these

D. True or False

1 to 12 all are true

E. Fill in the Blanks

Sl.No.	Answer	Sl.No.	Answer
1.	Computer	7.	Database
2.	15	8.	Database
3.	*Sui-generis*	9.	1998
4.	Compulsory	10.	Automatically
5.	Database	11.	Sale license
6.	Database	12.	Infringement

F. Short or One Word Answer

Sl.No.	Answer	Sl.No.	Answer
1.	Data	9.	Data type
2.	Database	10.	Extraction
3.	Database rights	11.	Reutilization
4.	Data management	12.	DBMS
5.	Data definition language	13.	Meta data
6.	Data manipulation language	14.	Occurrence
7.	Data model	15.	Open data connectivity
8.	Data normalization		

CHAPTER 15

Mask Work Rights

A. Definitions

Integrated Circuits: The final or intermediate form of any product (i) having two or more layers of metallic, insulating, or semiconductor material, deposited or otherwise placed on or etched away or otherwise removed from a piece of semiconductor material in accordance with a predetermined pattern; and (ii) intended to perform electronic circuitry functions.

Mask Work: The layout is called a mask work because, in photolithographic processes, the multiple etched layers within actual ICs are each created using a mask, called the photo mask, to permit or block the light at specific locations, sometimes for hundreds of chips on a wafer simultaneously.

Mask Work Rights: The intellectual property right conferring time-limited exclusivity to reproduction of a particular layout.

B. Background Information

1. A **mask work** is a two or three-dimensional layout or topography of an integrated circuit (IC or "chip"), *i.e.* the arrangement on a chip of semiconductor devices such as transistors and passive electronic components such as resistors and interconnections.

2. Mask work protection is protected by *sui-generis* rights.

3. The correct symbol used in a mask work's copyright notice is Ⓜ (**M** enclosed in a circle;

4. Mask work rights are closer to copyrights than with other exclusive rights such as patents or trademarks.

5. They are used alongside copyright to protect a read-only memory (ROM) component that is encoded to contain computer software.

6. **Registration:** The registration is not compulsory for mask works. But registration is essential to get legal rights.The registration process consists of the following steps:

7. **Filing Application:** The application has to be filed in the Patents and Copyrights office in the prescribed form which cab be obtained from the copyrights office. The application for registration may be submitted by the owner of the mask work or an agent duly authorized by the owner.

8. **Fee:** The prescribed processing fee has to be deposited along with application form.

9. **Declaration Certificate:** A declaration has to be submitted stating that the applicant is authorized to submit the application and the statements made are correct to the best of knowledge of the person filing the application.

   The application should name and signature [original] of the applicant.

   For the same version of a mask work only one registration can be made.

10. Mask works must be registered within 2 years of first commercial exploitation. The date of registration is the date on which a complete application, fee, and an appropriate deposit representing an acceptable filing are filed with the Copyright Office. If the Mask Work is not registered within this time period, protection under the SCPA will be lost.

11. **Matters Covered:** Mask work rights cover two or three-dimensional layout or topography of an integrated circuit (IC or "chip"), i.e. the arrangement on a chip of semiconductor devices such as transistors and passive electronic components such as resistors and interconnections.

12. **Duration of Protection:** The rights in semiconductor mask work last only two years (if unregistered) or ten years (if registered). The SCPA provides protection of a Mask Work for a period of 10 years from the date of first commercial exploitation, or the date of registration with the Copyright Office, whichever occurs first.

13. **Rights Covered:** During the term of protection, the owner of a mask work has the following exclusive rights:

    (*i*) To reproduce the mask work by optical, electronic, or any other means;

    (*ii*) To import or distribute a semiconductor chip product in which the mask work is embodied;

    (*iii*) To induce or knowingly to cause another person to do any of the acts described in number 1 and number 2.

    The exclusive rights in a mask work are somewhat like those of copyright: the right to reproduce the mask work or (initially) distribute an IC made using the mask work. A owner of a mask work may freely import, distribute or use, but not reproduce the chip (or the mask).

14. **Requirements:** The legal protection is provided to the three-dimensional images or patterns formed on or in the layers of metallic, insulating, or semiconductor material and fixed in a semiconductor chip product, i.e., the "topography" of the "chip." For protection, a mask work must be original. A mask work is considered "original" if it is the independent creation of an author who did not copy it.

15. **Authorization/Transfer:** The Act allows the owner of the exclusive rights to transfer all of them or to license all or fewer than all the rights. A transfer or license must be in writing and signed by the owner of the rights or by a duly authorized agent of the owner. The Act also provides that the exclusive rights in a mask work may be transferred by operation of law or by terms of a will; or, the rights may pass as personal property under the applicable state Law. In other words, the owner of amask work can transfer all the exclusive rights of his mask work to any other person.The new owner should obtained by transfer the totality of the rights in the mask work.

16. **Enforcement:** Protection for a mask work begins on the date the mask work is registered with the Copyright Office, or the date the mask work is first commercially exploited anywhere in the world, whichever occurs first. Protection lasts for 10 years.The protection expires at the end of the tenth calendar year after it began.

    The mask work rights come into force immediately after creation of mask work. In other words, mask work rights exist automatically [like copyrights] when they are created, regardless of registration. In case of patents, rights are confered only after registration.

17. **Infringement:** After a certificate has been issued by the Copyright Office, the owner of the mask work or the exclusive licensee of all rights in the mask work may institute a civil action for infringement occurring after the commencement of protection accorded by the Act.

18. **Notice and Symbol:** An appropriate notice on a semiconductor chip or product includes either the words "Mask Work," the symbol *M*, or the symbol Ⓜ, followed by the name of the owner or a recognizable abbreviation of the owner's name. Such notice is optional, but provides certain benefits to the owner with respect to enforcing rights.

19. Certain activities by competitors are permitted, such as reverse engineering for the purposes of teaching, analyzing, or evaluating the concepts embodied in the mask work including the circuitry, logic flow, and organization of components.

20. The results of legitimate reverse engineering may be incorporated into another original mask work and commercially exploited.

21. Any idea or concept, process, principle, discovery or system associated with [a semiconductor product] a mask work is not protected.

C. Multiple Choice Questions

1. ***Sui-generis* rights include:**
   (a) Database rights (b) Mask work rights
   (c) Moral rights (d) All of these

2. **Mask work comes under:**
   (a) Primary rights (b) Ordinary rights
   (c) *Sui-generis* rights (d) All of these

3. **Unregistered mask work last for:**
   (a) 2 years (b) 5 years
   (c) 7 years (d) 10 years

4. **Registered mask work last for:**
   (a) 2 years (b) 5 years
   (c) 7 years (d) 10 years

5. **Mask work includes:**
   (a) Transistors (b) Resistors
   (c) Inter connections (d) All of these

6. **A mask work permits analysis of:**
   (a) Circuitry (b) Logic flow
   (c) Components (d) All of these

7. **Mask work is closer to:**
   (a) Copyrights (b) Patent
   (c) Trade mark (d) Trade secret

8. **In USA mask work rights were first introduced in:**
   (a) 1950 (b) 1970
   (c) 1984 (d) 1994

9. **Mask work rights were first introduced in:**
   (a) England (b) Germany
   (c) USA (d) Japan

D. True or False

1. Mask work is related to IC chips.
2. Unregistered mask work lasts for 2 years.
3. Registered mask work lasts for 10 years.
4. Mask work rights come under *sui-generis* rights.
5. For protection a mask work must be original.
6. Protection of mask work begins immediately after registration.
7. Mask work rights are closer to copyrights.
8. Mask work rights should be registered within 2 years of first commercial exploitation.
9. Transfer of mask work rights to other person is permitted.
10. Mask work rights are liable to infringement.

E. Fill in the Blanks

1. Mask work is related to ________.
2. Unregistered mask work lasts for ________ years.
3. Registered mask work lasts for ________ years.
4. Mask work rights come under ________ rights.
5. For protection a mask work must be ________.
6. Protection of mask work begins immediately after ________.
7. Mask work rights are closer to ________.
8. Mask work rights should be registered within ________ years of first commercial exploitation.
9. Transfer of mask work rights to other person is ________.
10. Mask work rights are liable to ________.

F. Short or One Word Answer

1. A two or three dimensional layout or topography of an IC or integrated Chip
2. The intellectual property right conferring time-limited exclusivity to reproduction of a particular layout.

3. The final or intermediate form of any product (1) having two or more layers of metallic, insulating, or semiconductor material, deposited or otherwise placed on or etched away or otherwise removed from a piece of semiconductor material in accordance with a predetermined pattern; and (2) intended to perform electronic circuitry functions.

ANSWERS

C. Multiple Choice Questions

Sl.No.	*Answer*	*Sl.No.*	*Answer*
1.	(d) All of these	**6.**	(d) All of these
2.	(c) *Sui-generis* rights	**7.**	(a) Copyrights
3.	(a) 2 years	**8.**	(c) 1984
4.	(d) 10 years	**9.**	(c) USA
5.	(d) All of these		

D. True or False

1 to 10 all are true

E. Fill in the Blanks

Sl.No.	*Answer*	*Sl.No.*	*Answer*
1.	IC Chip	**6.**	Registration
2.	2	**7.**	Copyrights
3.	10	**8.**	2
4.	*Sui-generis*	**9.**	Permitted
5.	Original	**10.**	Infringement

F. Short or One Word Answer

Sl.No.	*Answer*	*Sl.No.*	*Answer*
1.	Mask work	**3.**	Integrated circuits
2.	Mask work rights		

Chapter 16

Plant Breeders' Rights

A. Definitions

Breeders' Exemption: Rights provided to use protected material as the basis to develop a new variety or for other research work; also called research exemptions or breeders' privilege.

Distinctiveness: The variety must be distinguishable in atleast one character from previously available varieties.

Farmers' Exemption: Legal rights provided to farmers to save, use, exchange, share or sell his farm produce of a protected variety; also called farmers' rights or farmers' privilege.

Novelty: The newness of a variety means not previously sold,

Plant Breeders' Rights: Plant breeders' rights [PBR], also known as plant variety rights (PVR), are intellectual property rights granted to the breeder of a new variety of plant.

Stability: Each generation of the variety should be similar.

Uniformity: All plants in a variety should look alike.

B. Background Information

1. Plant breeders' rights [PBR], also known as plant variety rights (PVR), are intellectual property rights granted to the breeder of a new variety of plant.

2. Plant Breeders Rights are granted to novel plant varieties that are distinctive, uniform, and stable (*e.g.*, cultivars breed true-to-type for desired traits).

3. The legal protection of a new plant variety is granted to the breeder or his successor. The effect of PBR is that prior authorization is required before the material can be used for commercial purposes.

4. **Registration:** Registration is essential to get legal rights. There is no protection without registration.

5. **Duration:** The period of protection varies with plant species. The protection is provided for a definite period. The duration of protection for different plant species is as follows'

   (*i*) For Trees and Vines: Initially for nine years and maximum for 18 years.

   (*ii*) For extant varieties: Initially six years and maximum for 15 years fro the date registration of the variety.

   (*iii*) In other cases: Initially six years and maximum for 15 years fro the date registration of the variety.

6. Thus breeders' rights allow plant breeders to get benefit of their variety for a limited period. When the PBR period is over, the variety reverts to the public domain and is available to every body. The limited duration of Plant Breeders' Rights ensure a balance between private and public interest.

7. **Validity:** The protection of the Plant Breeders' Rights is valid only in the country where it has been registered. The protection in other countries can be obtained by filling separate application in each country.

8. **Matters Covered:** The protection right can be granted for varieties of all botanical genera and species. The variety should have a designation [name] as per the rule of International Code of Nomenclature.

9. **Requirements:** There are four basic requirements for protection of a variety under PBR, *viz.*, (i) novelty, (ii) distinctiveness, (iii) uniformity, and (iv) stability. See later.

10. **Transfer:** An authorized plant breeder has right to authorize other interested persons for commercial production and marketing of his variety on royalty basis.

11. **Control:** The authorized plant breeder has right to prevent other from commercial production and marketing of his variety without permission.

12. **Enforcement:** The Plant Breeders' Rights come into force immediately after registration of the variety.

13. **Other Points:** There are some other points which are taken into account for granting PBR.

14. BPR can be granted only to breeder or legal owner of a new variety.

15. The plant breeders rights protect the variety but not the standard breeding procedures that are used to develop new variety.

16. The main purpose of PBR is to encourage breeders to develop better plant varieties.

17. The PBRs provide monopoly of new varieties to breeders for a limited period.

18. The same variety can be patented and protected by PBR. The gene technology or gene sequences can be patented and the end product that is the plant variety can be protected under PBR.

19. A person growing protected variety can not sale its seed on commercial scale. But he can use the produce of such seed for planting purpose on his farm.

20. PBR coexists with other legislation. It is not independent of other laws of the land.

21. The protected variety can be used to breed new varieties.

22. A variety cannot be registered under trademarks. It comes under Sui generic rights, hence it can be registered under PBR only.

23. There are four basic requirements for protection of a variety under Plant Breeders' Rights. These are (i) novelty, (ii) distinctiveness, (iii) uniformity, and (iv) stability.

24. **Novelty:** It refers to newness of a variety. The variety should be new and should not have been commercially cultivated for more than one year before granting protection under PBR. In other words, there should not be a prior sale for more than one year.

25. **Distinctiveness:** The new variety must be distinguishable in one or more morphological, quality or other character from previously available varieties. When a variety differs atleast in one characteristic from varieties of common knowledge, it is known as distinct.

26. **Uniformity:** The variety should be uniform. In other words, all plants in a variety should look alike.

27. **Stability:** The should give stable performance in different generations. All these attributes are determined by grow out test.

28. The Plant Breeders' Rights provide exclusive rights to the breeder for commercial production and marketing of his variety.

29. Unauthorized production and marketing of a registered variety by other person amounts to infringement. The owner has the right to take legal

action against the infringer and claim damages. The PBR Authority may initiate legal action against a person who is involved in the infringement of a protected plant variety. The Authority can recover both damages and profits from such person.

30. The Plant Breeders' Act [PBRA] provides for heavy penalties against infringement of the breeders' right. In USA it 55,00 Dollars for individuals and 275000 Dollars for companies.

31. The Plant Breeders' Rights Act provides two exemptions, *viz.*, (i) breeders' exemptions and (ii) farmers' exemptions.

C. Multiple Choice Questions

1. **Plant breeders' rights come under:**
   (a) Primary rights (b) Ordinary rights
   (c) *Sui-generis* rights (d) All of these

2. **Maximum protection period for field crops is:**
   (a) 6 years (b) 10 years
   (c) 15 years (d) 20 years

3. **Minimum protection period for field crops is:**
   (a) 6 years (b) 9 years
   (c) 12 years (d) 15 years

4. **Maximum protection period for trees and vines is:**
   (a) 9 years (b) 10 years
   (c) 18 years (d) 20 years

5. **Minimum protection period for trees and vines is:**
   (a) 9 years (b) 10 years
   (c) 18 years (d) 20 years

6. **New variety for protection must have:**
   (a) Distinctiveness (b) Uniformity
   (c) Stability (d) All of these

7. **PBR may lead to:**
   (a) Malpractices (b) Monopoly
   (c) Increase in price (d) All of these

8. **First UPOV Act was introduced in:**
   (a) 1968 (b) 1975
   (c) 1978 (d) 1982

9. Second UPOV Act was introduced in:

(a) 1968 (b) 1975

(c) 1991 (d) 2001

10. PBR may lead to reduction in:

(a) Seed price (b) Biodiversity

(c) Seed quality (d) All of these

D. True or False

1. Plant breeders' rights come under *sui-generis* rights.
2. For Plant breeders' rights registration is compulsory.
3. Plant breeders' rights come into force immediately after registration.
4. Plant breeders' rights are valid only in the country of registration.
5. Transfer of PBR to other persons is permitted on royalty basis.
6. New varieties of trees and vines are initially protected for 9 years.
7. Maximum protection period for trees and vine varieties is 18 years.
8. New varieties of field crops are initially protected for 6 years.
9. Maximum protection period for new varieties of field crops is 15 years.
10. The first UPOV Act came into force in 1978.
11. The second UPOV Act was introduced in 1991.
12. PBRs are liable to infringement.

E. Fill in the Blanks

1. Plant breeders' rights come under ________ rights.
2. For Plant breeders' rights registration is ________.
3. Plant breeders' rights come into force immediately after ________.
4. Plant breeders' rights are valid only in the country of ________.
5. Transfer of PBR to other persons is permitted on ________ basis.
6. New varieties of trees and vines are initially protected for ________ years.
7. Maximum protection period for trees and vine varieties is ________ years.
8. New varieties of field crops are initially protected for ________ years.

9. Maximum protection period for new varieties of field crops is ________ years.

10. The first UPOV Act came into force in ________.

11. The second UPOV Act was introduced in ________.

12. PBRs are liable to ________.

F. Answer in Short or One Word

1. Intellectual property rights granted to the breeder of a new variety of plant.

2. Rights provided to to plant breeders to use protected material as the basis to develop a new variety or for other research work.

3. Legal rights provided to farmers to save, use, exchange, share or sell his farm produce of a protected variety.

4. The newness of a variety means not previously sold

5. A variety which distinguishable in at least one character from previously available varieties.

6. All plants in a variety should look alike.

7. Similar performance of a variety in repeated cultivation.

ANSWERS

C. Multiple Choice Questions

Sl.No.	*Answer*	*Sl.No.*	*Answer*
1.	(c) *Sui-generis* rights	**6.**	(d) All of these
2.	(c) 15 years	**7.**	(d) All of these
3.	(a) 6 years	**8.**	(c) 1978
4.	(c) 18 years	**9.**	(c) 1991
5.	(a) 9 years	**10.**	(b) Biodiversity

D. True or False

1 to 12 all are true

E. Fill in the Blanks

Sl.No.	*Answer*	*Sl.No.*	*Answer*
1.	*Sui-generis*	**7.**	18
2.	Compulsory	**8.**	6
3.	Registration	**9.**	15
4.	Registration	**10.**	1978
5.	Royalty	**11.**	1991
6.	9	**12.**	Infringement

F. Short or one word Answer

Sl.No.	*Answer*	*Sl.No.*	*Answer*
1.	Plant breeders' rights	**5.**	Distinctiveness
2.	Breeders' exemptions	**6.**	Uniformity
3.	Farmers' exemptions	**7.**	Stability
4.	Novelty		

Chapter 17

Farmers' Rights

A. Definitions

Biological Diversity: Total diversity within species, between species and of ecosystems."

Biotechnology: Biotechnology" means "any technological application that uses biological systems, living organisms, or derivatives thereof, to make or modify products or processes for specific use."

Cultivar: Cultivated variety of plants.

Domestic biodiversity: the genetic variation existing among the species, breeds, cultivars and individuals of animal, plant and microbial species that have been domesticated, often including their immediate wild relatives.

***Ex-situ* conservation**: the conservation of components of biological diversity outside their natural habitats, often in a laboratory, collection, botanical garden, zoo, or aquarium.

Farmers' rights: rights arising from the past, present and future contributions of farmers in conserving, improving and making available plant or animal genetic resources.

Genetic Resources: Genetic resources means genetic material of actual or potential value

In-situ: in the original location. (Opposite: ex-situ).

In-situ **conditions**: The conditions where genetic resources exist within ecosystems and natural habitats; in the case of domesticated or cultivated species, in the surroundings where they have developed their distinctive properties.

Landrace: a crop cultivar or animal breed that evolved with and has been genetically improved by traditional agriculturists, but has not been influenced by modern breeding practices.

Public Domain: Public Domain refers to the knowledge that is freely available, commonly shared throughout the world without any access restrictions. It cannot be protected by intellectual property rights.

B. Background Information

1. Farmers' rights refer to the rights arising from the past, present and future contributions of farmers in conserving, improving and making available plant or animal genetic resources, particularly those in the centres of origin/diversity.

2. It refers to the legal rights provided to farmers to save, use, sow, replant, exchange, share or sell his farm produce including seed of a variety protected under Plant Variety Protection Act.

3. The purpose of these rights is to "ensure full benefits to farmers and support the continuation of their contributions.

4. Farmers' rights ensure global recognition of the need for conservation and the availability of sufficient funds for these purposes;

5. Farmers' rights assist farmers and farming communities throughout the world, especially those in areas of original diversity of plant genetic resources, in the protection and conservation of their PGR and of the natural biosphere; and,

6. Farmers' rights allow the full participation of farmers, their communities and countries in the benefits derived, at present and in the future, from the improved use of PGR.

7. **Registration:** A farmer who has bred or developed a new variety shall be entitled for registration and protection of his variety like a breeder.

8. **Requirements:** The farmers; variety shall be entitled for registration if it meets the following requirements.
   (*i*) It should have a denomination [name] assigned by the applicant
   (*ii*) An affidavit that the variety does not contain terminator gene.
   (*iii*) A complete passport data of parental lines from which the variety has been derived.
   (*iv*) The contribution of any farmer, village community, institution or organization in the development of the variety.
   (*v*) The source of genetic material used in developing the variety and the parental material used in developing new variety was lawfully acquired.
   (*vi*) A brief description of the variety about its novelty, distinctiveness, uniformity and stability.
   (*vii*) A proof of the right to make the application

9. **Exemption from Fee:** A farmer or group of farmers or village community shall not be liable to pay any fee in any proceeding before the Authority or Registrar or the Tribunal or the High Court under Plant Variety Protection and Farmers' Rights Act or the rules made there under.

10. **Rights Covered:** The Farmers' Rights Act provide following rights to the farmers in India even for the variety protected under PVP and FR Act.

    (i) Rights to save, use, sow or replant seeds of his farm produce.

    (ii) Rights to exchange or share his farm produce, and

    (iii) Rights to sell seeds of his farm produce

    However, the farmer cannot sell the seeds or planting material under a brand name.

11. **Rights of Communities:** In India, any of the following can file application for registration on behalf of any village or community.

    (*i*) Any person or group of persons [whether actively engaged in farming or not], and

    (*ii*) Any governmental or non-governmental organization

    After completing all formalities, if the Authority is satisfied, the variety can be registered. If due credit has not been given to the farmers or communities who have helped in the development of the variety, the breeder has to pay compensation to the concerned party. Or part of compensation will be deposited in the gene fund.

12. **Authorization of Farmers' Variety:** The authorization for an essentially derived variety from farmers' variety cannot be given by the breeder of such farmers; variety except with the consent of the farmers or group of farmers or community of farmers who have made contribution in the preservation or development of such variety.

13. **Materials Covered:** Farmers' Rights include new varieties of all self pollinated, cross pollinated and asexually propagated varieties.

14. **Duration:** The protection is provided for a definite period. The duration of protection for different plant species is as follows'

    (i) For Trees and Vines: Initially for nine years and maximum for 18 years.

    (ii) For extant varieties: Initially six years and maximum for 15 years fro the date registration of the variety.

    (iii) In other cases: Initially six years and maximum for 15 years fro the date registration of the variety.

15. **Validity:** The protection is valid only in the country where the variety has been registered.

16. **Benefit Sharing:** If the variety has been developed by a particular farmer, he can get benefits of his variety. In case of community rights, the sharing of benefits becomes difficult and the money is deposited in the gene fund.

17. **Gene Fund:** The central government may constitute a Fund to be called National Gene Fund. The money earned from following will be deposited in this fund;

    (i) The benefit sharing received from the breeder of a variety

    (ii) The annual payable to the Authority in the form of royalty.

    (iii) The amount of compensation received by the Authority.

    (iv) Contribution received from amy National or International Organization and other sources.

    The gene fund may be utilized for benefit sharing, rewards and other expenditure related to Plant Variety Protection and Farmers' Rights Act.

18. **Rewards:** A farmer who is engaged in the conservation of genetic resources of land races and wild relatives of economic plants and their improvement through selection shall be entitled for recognition and reward. The expenses toward rewards will be met out from the National Gene Fund.

19. **Acknowledgement by Breeders:** It is compulsory for the breeder of a variety to reveal the origin of the parental lines plant genetic resources used in the development of a variety. It will help in deciding the share of different persons, *viz.*, farmers and other communities in benefits earned by breeder of such variety.

20. **Infringement:** If a farmer was not aware of the PVP and FR Act at the time of using any protected variety, he will be exempt from the infringement and will be asked not to do such act in future. This right has been provided to Indian farmers by PVP and FR Act 2001.

21. **Ancient Rights:** Farmers' Rights have a deep historic character. They existed since humans created agriculture to serve their needs, and played vital role in conservation of biodiversity. Farmers are the guardians of these genetic resources, which support the evolution of species. Farmers are the inheritors of the skills and knowledge of the generations that have created this biological wealth. For these contributions, it is necessary to recognize Farmers' Rights.

22. **Rights Covered:** Farmers' Rights include the right over resources and associated knowledge, united indivisibly, and mean the acceptance of traditional knowledge, respect for cultures and recognition that these are the basis of the creation of knowledge.

23. **Property Rights:** The farmers deserve right to control, the right to decide the future of genetic resources, the rights to define the legal framework of property rights of these resources.

24. **Collective Nature:** Farmers' Rights are of an eminently collective nature and therefore are protected by *sui-generic* rights.

25. **Application:** These rights should have a national application, respecting the sovereignty of each country, to establish local laws based on these principles.

26. **Rights to Means:** It includes territorial rights, right to land, right to water and air and rights to conserve biodiversity and achieve food security.

27. **Rights for Active Participation:** The farmers deserve right to participation in the definition, elaboration and execution of policies and programmes linked to genetic resources.

28. **Rights to Technology:** The farmers deserve right to appropriate technology as well as participation in the design and management of research programmes.

29. **Rights to Benefit Sharing:** The farmers deserve right to define the control and handling of benefits derived from the use, conservation and management of these resources.

30. **Rights on Genetic Resources:** The farmers deserve right to use, choose, store and freely exchange of genetic resources.

31. **Rights for Sustainable Agriculture:** The farmers deserveright to develop models of sustainable agriculture that protects the biodiversity and to influence the policies that support it.

32. The purpose of farmers' rights is to encourage farmers and farming communities to nurture and conserve, and to utilize and improve, plant genetic resources (FAO, 1996).

    Farmers' Rights are generally considered to be group (collective) rights that do not necessarily imply direct compensation or bilateral reciprocity, such as selling a license for exclusive rights to collect and use crop germplasm.

33. Farmers both men and women have conserved plant genetic resources over the past thousands of years.

34. Farmers are rightly called as on site managers of plant genetic resources.

35. Farmers have in depth knowledge of the land races and wild relatives of cultivated plant species.

36. All modern plant varieties contain only those genes that have originated from farmers' traditional varieties (landraces) or wild crop relatives

37. The women and men have historically provided the germplasm upon which scientific plant breeding is based.

38. Generally, germplasm is provided by small-scale subsistence farmers to plant breeders whose varieties are then adapted/developed for medium to large scale commercial farmers (Spillane, 1996).

39. Women's special knowledge of the value and diverse uses of plants for nutrition, health and income has important implications for the conservation of plant genetic resources.

40. The women have played significant role not only in terms of labour and skills, but also their decision-making about how natural resources are used to satisfy the multiple needs of rural households.

41. In much of the developing world, the on-farm and in-situ conservation and use of plant genetic resources begins with women. As farmers, they are responsible for growing crop plants and collecting seed for next year sowing.

42. It is estimated that ninety per cent of the planting materials used in developing countries is produced by the farmers themselves, *i.e.* from saved seed (Spillane, 1996). 43. The selection of wild genetic resources for home planting, of seeds for conserving for next year's planting and of asexually propagated material is usually carried out by women. This selection is a sophisticated process that takes into account many different criteria such as taste, color, palatability and texture; resistance to pests and diseases; adaptation to soil and agro-climatic conditions and so on. The selection criteria are built up over years of experience.

44. Farmers in all regions of the world should be assisted in their conservation of natural resources and more specifically the selection, conservation, improvement and sustainable use of PGR.

45. Non Governmental Organizations should assist in the entire process as they are close to the communities and have the capacities to provide long-term support. The establishment of strong village level institutions with adequate representation of both women and men farmers can be key in promoting successful programmes.

46. Incentives should be used to enable farmers to divert land, labour and capital towards conserving biological resources and to facilitate the participation of certain groups in activities that promote the sustainable use of those resources.

47. The participation of women farmers and the recognition of their contributions in the design and elaboration of such policies and programmes at all levels is essential.

48. Those farmers who develop breeding materials should be empowered to decide, at their discretion, if the materials will be made available to other (*i.e.* external) users.

49. There should be equal rights for conservation and participation in PGR related programs and policies, which is often lacking in developing countries.

50. Participatory research and plant breeding should be promoted with different socio-economic groups in rural communities in order to improve the productivity of local varieties and to promote marketing of special products derived from these materials.

51. Modern plant breeding has concentrated on developing a small number of varieties that are widely adapted to many environments. Land races should also be used for developing varieties suitable for diverse ecological conditions.

52. There is need for the creation of an Intentional Fund to support in-situ conservation and to benefit rural men and women who produce and conserve PGR. In general farmers are interested in conserving resources but lack of financial support acts as a barrier in their way.

53. PGR conservation should be integrated into other community development programs It will promote and employment in activities related to PGR building up of local institutions, capacity building) and collecting and dissemination;, information on the status and trends of PGR.

54. There should be equitable sharing of benefits arising from the utilization of PGR, knowledge, innovations and practices.

C. Multiple Choice Questions

1. **Which of the following comes under *sui-generis* rights?**

   (a) Patent (b) Trademark

   (c) Farmers' rights (d) Industrial designs

2. **Farmers' rights come under:**

   (a) Primary rights (b) Secondary rights

   (c) *Sui-generis* rights (d) None of these

3. **Farmers conserve PGR in their:**

   (a) Field (b) Garden

   (c) Kitchen gardens (d) All of these

4. **Which of the following comes under *sui-generis* rights?**

   (a) Patent (b) Trademark

   (c) Farmers' rights (d) Industrial designs

5. **Farmers conserve PGR in the form of:**
   (a) Landraces (b) Wild species
   (c) Wild relatives (d) All of these

6. **Which of the following is of collective nature?**
   (a) Breeders' rights (b) Farmers' rights
   (c) Copyrights (d) Moral rights

7. **Farmers rights were introduced by FAO in:**
   (a) 1979 (b) 1989
   (c) 1999 (d) 2001

8. **For field crops, farmers' rights are protected maximum for:**
   (a) 6 years (b) 10 years
   (c) 15 years (d) 20 years

9. **For trees and vines, farmers' rights are protected maximum for:**
   (a) 9 years (b) 10 years
   (c) 18 years (d) 20 years

10. **For field crops, minimum period of farmers' rights protection is:**
    (a) 6 years (b) 9 years
    (c) 12 years (d) 15 years

D. True or False

1. Farmers' rights come under *sui-generis* rights.
2. Farmers rights are of collective nature.
3. Land races are also known as primitive cultivars.
4. New varieties of trees and vines are initially protected for 9 years.
5. Maximum protection period for trees and vine varieties is 18 years.
6. New varieties of field crops are initially protected for 6 years.
7. Maximum protection period for new varieties of field crops is 15 years.
8. Farmers' rights came into force in 1989.
9. On site manager of PGR refers to farmers.
10. Both man and women were responsible for conservation of PGR in the past.

E. Fill in the Blanks

1. Farmers' rights come under ________ rights.
2. Farmers rights are of ________ nature.
3. Land races are also known as ________ cultivars.
4. New varieties of trees and vines are initially protected for ________ years.
5. Maximum protection period for trees and vine varieties is ________ years.
6. New varieties of field crops are initially protected for ________ years.
7. Maximum protection period for new varieties of field crops is ________ years.
8. Farmers' rights came into force in ________.
9. On site manager of PGR refers to ________.
10. Both man and women were responsible for conservation of ________ in the past.

F. Answer in Short or One Word

1. Variety of alleles found in living organisms.
2. Application of living organisms, for commercial production of products or processes for specific use.
3. Cultivated variety of plants.
4. Variety of alleles found in domesticated plants, animals and micro-organisms.
5. Conservation of biological diversity outside their natural habitats.
6. Rights arising from the past, present and future contributions of farmers in conserving, improving and making available plant or animal genetic resources.
7. Genetic material of actual or potential value
8. Conservation in the original location.
9. The conditions where genetic resources exist within ecosystems and natural habitats.
10. A primitive cultivar generally developed by farmers.
11. The knowledge that is freely available, commonly shared throughout the world without any access restrictions.

ANSWERS

C. Multiple Choice Questions

Sl.No.	*Answer*	*Sl.No.*	*Answer*
1.	(c) Farmers' rights	**6.**	(b) Farmers' rights
2.	(c) *Sui-generis* rights	**7.**	(b) 1989
3.	(d) All of these	**8.**	(c) 15 years
4.	(c) Farmers' rights	**9.**	(c) 18 years
5.	(d) All of these	**10.**	(a) 6 years

D. True or False

1 to 10 all are true

E. Fill in the Blanks

Sl.No.	*Answer*	*Sl.No.*	*Answer*
1.	*Sui-generis*	**6.**	6
2.	Collective	**7.**	15
3.	Primitive	**8.**	1889
4.	9	**9.**	Farmers
5.	18	**10.**	PGR

F. Short or One Word Answer

Sl.No.	*Answer*	*Sl.No.*	*Answer*
1.	Biological diversity	**7.**	Genetic resources
2.	Biotechnology	**8.**	*In-situ* conservation
3.	Cultivar	**9.**	*In-situ* conditions
4.	Domestic diversity	**10.**	Landrace
5.	Ex-situ conservation	**11.**	Public domain
6.	Farmers rights		

Chapter 18

Moral Rights

A. Definitions

Disclosure right: The right to withhold publication of an image or broadcast.

False attribution Right: The right not to have a work falsely attributed to them.

Integrity Right: The right to prevent derogatory treatment of their work.

Moral Rights: Moral rights are a special extension to copyright that gives the originator of a copyrighted work rights over its use.

Paternity Right: the right to be acknowledged as the author.

B. Background Information

1. Moral rights are rights of creators of copyrighted works generally recognized in civil law jurisdictions and first recognized in France and Germany, before they were included in the Berne Convention for the Protection of Literary and Artistic Works in 1989.

2. Moral rights are a special extension to copyright that gives the originator of a copyrighted work rights over its use. The originator can specify moral conditions on the use or exploitation of their creation, even when the rights are licensed or sold to others'.

3. The term "moral rights" is a translation of the French term "droit moral," and refers not to "morals" as advocated by the religious right, but rather to the ability of authors to control the eventual fate of their works. An author is said to have the "moral right" to control his or her work. The concept of moral rights thus relies on the connection between an author and his/her creation. Authors may seek moral rights protection from state moral rights law.

4. Moral rights protection varies around the world. For example, in the United States, moral rights protection is at a lower level that in many European Union countries. France, by way of comparison, has one of the most protective moral rights regimes in the world.

5. Like registration is not compulsory for moral rights. But it is desirable to register moral rights to have legal control over the unauthorised use of moral rights.

6. All those matters which are covered under copyrights are also covered by moral rights.

7. The term of moral rights is the same as the term of copyright protection. So when copyright protection ends and a work falls into the public domain the moral rights protection ceases at the same time.

8. Moral rights are not transferrable, and end only with the life of the author, whereas copyrights continue up to 70 years after the death of the author.

9. Moral rights are separate rights to that of copyright and belong to the author of a work. The four categories of moral rights, *viz.* right to paternity, right to integrity, right to attribution and right of privacy are associated with authorship, and are quite distinct from copyright.

10. Since moral rights are linked with copyrights, there are no special requirements for protection of moral rights.

11. Moral rights related to copyright work are considered personal to the author and that cannot therefore be bought, sold or transferred.

12. The moral rights cannot be assigned, but can be waived contractually. Under Canadian copyright law a creator cannot assign moral rights.

13. It is only possible for a creator to agree not to exercise moral rights by "waiving" the rights. Authors may, however, waive their moral rights if do so in writing.

14. Moral rights can be waived for one particular individual or for one particular use.

15. The moral rights become effective immediately after release of the copyrighted works.

16. The determination of infringement involves two types of components, *viz.* a subjective and an objective component. The subjective component is whether the author himself feels that his work is used in a manner which is prejudicial to his honour or reputation. The object component is an evaluation of public or expert opinion.

17. The word of the author alone is not enough to establish infringement. To sustain a claim of infringement, an author must be supported by expert or public opinion. There are various mechanisms with which an author may enforce his moral rights.

18. An author may show that, in altering or distorting her work, someone has created a "derivative work," thereby violating the Copyright Act.

19. If authorship of a work is attributed to an author against her will, or misattributed, the author may have a state action for defamation against the person responsible for the attribution.

20. If a person uses the identity of an author, or the works of the author, for her own benefit without the author's permission, then she may have violated the author's right of publicity or may be guilty of misappropriation of the author's work.

21. The person entitled to any of the moral rights can take legal proceedings in the civil courts for preventing infringements. He can ask for financial compensation for any damage caused by infringer including cost of legal proceedings.

C. Multiple Choice Questions

1. **Moral Rights include:**
   (a) Right of paternity (b) Right of integrity
   (c) Right to attribution (d) All of these

2. **Berne Convention granted moral rights to authors in:**
   (a) 1969 (b) 1974
   (c) 1989 (d) 1999

3. **The owner of moral rights has right to object to any:**
   (a) Distortion (b) Mutilation
   (c) Modification (d) All of these

4. **Moral rights cannot be:**
   (a) Bought (b) Sold
   (c) Transferred. (d) All of these

5. **Moral rights are valid till:**
   (a) 20 years (b) 50 years
   (c) 70 years (d) Author is alive

D. True or False

1. Moral rights are provided beyond those strictly recognized in copyright law.
2. Moral Rights were granted to authors under the Berne Convention (1989).
3. The creator or author shall have the right to claim authorship of the work even after sale or transfer of the copyright.
4. Moral Right cannot be transferred or assigned.
5. Authors may seek moral rights protection from moral rights laws.
6. Moral rights can be waived for one particular individual or for one particular use.
7. The moral rights become effective immediately after release of the copyrighted works.
8. Moral rights related to copyright work are considered personal to the author and that cannot therefore be bought, sold or transferred.
9. Like coprights registration is not compulsory for moral rights.
10. The moral rights cannot be assigned, but can be waived contractually
11. All those matters which are covered under copyrights are also covered by moral rights.
12. Moral rights end only with the life of the author

E. Fill in the Blanks

1. Moral rights are provided beyond those strictly recognized in ________.
2. Moral Right ________ transferred or assigned
3. Moral Rights were granted to authors under the ________ Convention
4. Moral Rights were granted to authors under the Berne Convention in ________.
5. Moral rights are valid till the author is ________.
6. Moral rights are related to ________ work.
7. Moral rights are governed by ________ rights.
8. Moral Rights are granted to ________.
9. The right to be acknowledged as the author is calle as right to ________.

10. The right to prevent derogatory treatment of copyrighted work is known as right to ________.

F. Answer in Short or One Word

1. A special extension to copyright that gives the originator of a copyrighted work rights over its use
2. The right to be acknowledged as the author.
3. The right to prevent derogatory treatment of their work.
4. The right not to have a work falsely attributed to them.
5. The right to withhold publication of an image or broadcast

ANSWERS

C. Multiple Choice Questions

Sl.No.	*Answer*	*Sl.No.*	*Answer*
1.	d. All of these	**2.**	c. 1989
3.	d. All of these	**4.**	d. All of these
5.	d. Author is alive		

D. True or False

1 to 12 all are true

E. Fill in the Blanks

Sl.No.	*Answer*	*Sl.No.*	*Answer*
1.	Copyrights	**2.**	Cannot be
3.	Berne	**4.**	1989
5.	Alive	**6.**	Copyrights
7.	*Sui-generis*	**8.**	Authors
9.	Paternity	**10.**	Integrity

F. Short or One Word Answer

Sl.No.	*Answer*	*Sl.No.*	*Answer*
1.	Moral Rights	**2.**	Paternity Right
3.	Integrity Right	**4.**	False attribution Right
5.	Disclosure Right		

Chapter 19

Supplementary Protection Certificates

A. Definitions

Copyright: Copyright refers to a document which grants exclusive right to the author/creator to publish and sell literary or musical or artistic work.

Patent: It is a legal document issued by a federal government that grants exclusive rights for the production, sale and profit from the invention of a product or process for a specific period of time. Patents also grant the right to prevent others from copying the invention.

***Sui-generis*:** *Sui-generis* refers to things of their own kind or things with unique characteristics.

Supplementary Protection Certificate: The Supplementary Protection Certificate (SPC) is a *sui-generis* form of legal protection that extends protection conferred by a patent (called "basic patent") in respect of a medicinal or plant protection product.

B. Background Information

1. A supplementary protection certificate (SPC) is a *sui-generic* form of intellectual property right [legal protection] that extends the protection conferred by a patent [called basic patent] in respect of medicinal products, such as drugs, and plant protection products, such as insecticides, and herbicides.

2. A pharmaceutical or plant protection product must obtain: (i) regulatory approval, (ii) a marketing authorization, and (iii) or product license before its sale in the market.

3. The grant of such an official authorization does often take years because of lengthy testing procedure.

4. Due to the long term procedure for an authorization a patent proprietor may lose valuable time, in which he would have been able to market his product.

5. The SPC compensates for the delay in authorization.

6. The SPC comes into force only after the expiry of corresponding patent.

7. **Duration:** It has maximum life time of 5 years when granted after 10 years after filing date of a patent. The term would be less than 5 years if it is granted after 5 years from the date of filing the corresponding patent.

8. **Conditions of Issue:** It is issued against a patent generally immediately on expiry of the term of the patent. It cannot be issued before 5 years from filing date of corresponding patent.

9. A separate application is to be filed for registration in each country.

   It extends the protection conferred by the basic patent beyond the term of that patent but in respect of medicinal or plant protection products covered by authorization.

10. It does not extend the term of the patent itself.

    A separate certificate is issued for each product, even if they are protected by the same basic patent.

11. **Registration:** The registration is required which is issued on national basis. The application is to be submitted in the prescribed application form in the patent office.

12. If the application does not meet the prescribed conditions, the applicant is asked to eliminate the deficiencies within an appropriate term (about two months).

13. If the deficiencies are not eliminated within the term set, the application shall be rejected. Generally, the above mentioned term is extendable.

14. **Matters Covered:** In European Union member countries, a **supplementary protection certificate** (short certificate) is a *sui-generis*, patent-like, intellectual property right. This type of right is available for medicinal products such as drugs and plant protection products, such as insecticides, fungicides and herbicides. The certificate is granted for the products which constitute (i) the active ingradient or combination of active ingradients of a medicinal product or (ii) the active substance or combination of active substances of a plant protection product.

15. **Duration of Protection:** The maximum life-span of the Supplementary Protection Certificate is five years from its effective date. The term of the certificate may not exceed five years. The term of an SPC depends on the date of issuance of the first market authorisation.
    (*i*) If the first market authorisation is issued less than five years after the filing date of the corresponding patent, no SPC is granted.
    (*ii*) If the first market authorisation is issued more than five years but less than ten years after the filing date of the corresponding patent, an SPC is granted for a term corresponding to the period elapsed between the five-year point and the market authorisation issuance date.
    (*iii*) If the first market authorisation is issued more than ten years after the filing date of the corresponding patent, an SPC is granted for a five-year term.
    (*iv*) The application for the certificate has to be filed within six months of the date on which the patent is granted. The supplementary Protection Certificate extends the life-span of a basic patent up to 5 years.
16. **Rights:** A certificate may only be grantedto the holder of the basic patent or his successor in title.The certificate confers the same rights as conferred by the basic patent and is subject to the same limitations and the same obligations.
17. **Requirements:** The following conditions should be fulfilled for grant of a supplementary protection certificates.

    The basic patent protecting the product should in force,

    The product has not already the subject of a certificate,

    A valid authorization has been grantedto place the product on the market.

    The authorization should be the first authorization to place the product on the market.
18. **Fees:** The prescribed processing fee is deposited along with the application for a supplementary Protection Certificate. An annual fee is payable for each year of the effective period of the certificate. Any period of less than 12 months is considered as a whole year. For instance, an effective period of 3 years and 6 months will require the payment of 4 years' annual fees. Annual fees are paid for maintaining a supplementary protection certificate. The yearly fees are paid every year three months before the due date and up to six months after the due date. If the yearly fees are paid after the due date, then an additional fee of 20 per cent is due with the yearly fee. The annual fees for a certificate are to be paid from the first year.
19. **Transfer:** The owner of the supplementary protection certificate has the power to transfer the right of certificate to any other person if he so desires. In other words, the owner or inventor may transfer ownership

to any one. Employee often assign such rights to their employers as part of their employment contracts.

20. **Enforcement:** A supplementary protection certificate comes into force only after the expiry of a basic patent. In other words, it enters into force when the related patent expires. The followings are also required before enforcement.

    (*i*) The basic patent has not previously lapsed or been revoked, and

    (*ii*) The annual fees are paid in time.

21. **Validity:** A Supplementary Protection Certificate does not give protection abroad; it is only effective in the country of registration. Applications for a supplementary protection certificate must be filed on a country-by-country basis. There is no unitary European supplementary protection certificate, but national ones only. The certificate will lapse under following situations.

    (*i*) at the end of the effective period,

    (*ii*) if the holder surrenders it,

    (*iii*) if the annual fees have not been paid in time, and

    (*iv*) if the product covered is no longer placed on the market.

    A lapsed certificate may be restored when a new authorization to place the product onn the market is granted. If a certificate is declared invalid, the fee for the remaining period will be refunded on written request by the holder.

22. **Infringement:** Like that of a basic patent, supplementary protection certificate provides legal rights against infringement of the protected materials.

    Documents provided in support of an application for a certificateincluding marketing authorizations, will normally be made open to public inspection after they are filed or sent to the office.Copies of any documents which are not treated as confidential will be available upon request as in case of documents relating to patents.

C. Multiple Choice Questions

1. **SPC comes under:**

    (a) Primary rights
    (b) Ordinary rights
    (c) *Sui-generis* rights
    (d) None of these

2. **The maximum life span of SPC is:**

    (a) 5 years
    (b) 10 years
    (c) 15 years
    (d) 20 years

3. **SPC is available for:**
   (a) Medicines (b) Insecticides
   (c) Pesticides (d) All of these
4. **SPC is related to:**
   (a) Copyrights (b) Trademarks
   (c) Basic patent (d) All of these
5. **SPC registration is issued on:**
   (a) Regional basis (b) National basis
   (c) International basis (d) All of these
6. **SPC is an extension of a:**
   (a) Copyright (b) Trademark
   (c) Patent (d) Trade name
7. **The fee of SPC is paid:**
   (a) Monthly (b) Quarterly
   (c) Half yearly (d) Yearly
8. **How much in advance SPC Fee can be paid?**
   (a) 1 month (b) 2 months
   (c) 3 months (d) 4 months
9. **How much after due date SPC Fee can be paid?**
   (a) 2 month (b) 3 months
   (c) 6 months (d) 8 months
10. **How time after basic patent, SPC can be issued?**
    (a) 3 years (b) 4 years
    (c) 5 years (d) 7 years

D. True or False

1. Supplementary Protection Certificate [SPC} rights come under *sui-generis* rights.
2. SPC is given to pharmaceutical or plant protection patent.
3. Maximum life of SPC is 5 years.
4. SPC can be issued only after 5 years of original patent.
5. SPC is an extension of a patent.

6. SPC comes into force only after expiry of original patent.

7. SPC extends the protection period of a patent.

8. SPC is valid only in the country of registration of a patent.

9. SPC is liable to infringement.

10. The owner of SPC has power to transfer it to any other person.

E. Fill in the Blanks

1. Supplementary Protection Certificate [SPC} rights come under ________ rights.

2. SPC is given to pharmaceutical or plant protection ________.

3. Maximum life of SPC is ________ years.

4. SPC can be issued only after ________ years of original patent.

5. SPC is an extension of a ________.

6. SPC comes into force only after ________ of original patent.

7. SPC extends the protection period of a ________

8. SPC is valid only in the country of ________ of a patent.

9. SPC is liable to ________.

10. The owner of SPC has power to ________ it to any other person.

F. Answer in Short or One Word

1. A *sui-generis* form of legal protection that extends protection of a basic a patent related to medicinal or plant protection products.

2. A legal document that grants inventor exclusive rights for the production, sale and profit from the invention of a product or process for a specific period of time.

3. Exclusive right granted to the author/creator to publish and sell literary or musical or artistic work.

4. Things of their own kind or things with unique characteristics.

ANSWERS

C. Multiple Choice Questions

Sl.No.	*Answer*	*Sl.No.*	*Answer*
1.	(c) *Sui-generis* rights	**6.**	(c) Patent
2.	(a) 5 years	**7.**	(d) Yearly
3.	(d) All of these	**8.**	(c) 3 months
4.	(c) Basic patent	**9.**	(c) 6 months
5.	(b) National basis	**10.**	(c) 5 years

D. True or False

1 to 10 all are true

E. Fill in the Blanks

Sl.No.	*Answer*	*Sl.No.*	*Answer*
1.	*Sui-generis*	**6.**	Expiry
2.	Patent	**7.**	Basic patent
3.	5	**8.**	Registration
4.	5	**9.**	Infringement
5.	Patent	**10.**	Transfer

F. Short or One Word Answer

Sl.No.	*Answer*	*Sl.No.*	*Answer*
1.	Supplementary protection certificate	**3.**	Copyright
2.	Patent	**4.**	*Sui-generis*

Chapter 20

Indigenous Property Rights

A. Definitions

Bio-Piracy: The unauthorized use of genetic resources and the knowledge associated with these genetic resources held by the communities living in the locality, by pharmaceutical companies. In other words, the illegal collection of indigenous plants by corporations who patent them for their own use.

Bio-prospecting: Bio-prospecting is the search for biological resources and related indigenous knowledge primarily for the purpose of commercial exploitation.

Community Knowledge: Knowledge held by communities, characterized by common or communal ownership.

Cultural Diversity: variety or multiple forms of human social structures, belief systems, and strategies for adapting to situations in different parts of the world.

Cultural Heritage: Monumental remains of cultures such as humankind, the dramatic arts, languages and traditional music, and the informational, spiritual and philosophical systems upon which creations are based.

Customary Law: Law that is customary. There have been requests that customary law may be given some recognition in the protection of traditional knowledge.

Expressions of Folklore: Folklore is a component of the broader term 'traditional knowledge'. 'Expressions of folklore' is often used in association with the system of copyright and related rights.

Intangible Heritage: The practices, representations, expressions, as well as the knowledge and skills, that communities, groups and, in some cases, individuals recognize as part of their cultural heritage.

Indigenous Knowledge: This term refers to knowledge held by communities and peoples that are indigenous. In comparison, traditional knowledge is a broader term as some traditional knowledge may not have the characteristic of indigenous knowledge.

Indigenous Peoples: People whose ancestors inhabited a place or country.

Intergenerational Equity: A core proposition is that future generations have a right to an inheritance (capital bequest) sufficient to allow them to generate a level of well-being no less than that of the current generation.

Non-Governmental Organization (NGO): A non-profit making, voluntary, service-oriented/development oriented organization.

Prior Informed Consent: Consent that is required for accessing genetic resources or the associated knowledge held by communities living in the locality; requires communities to be informed of all consequence of grant of such permission.

Scientific Knowledge: The knowledge gained through education and scientific readings.

Sui-generis: *Sui-generis* is a Latin phrase meaning of its own kind. It deals with issues such as Plant Breeders Rights, integrated circuits, moral rights, traditional knowledge, *etc.*

Sustainable Development: Development that meets the needs of the present without compromising the ability of future generations to meet their own needs.

Traditional Knowledge Holders: Refers to persons who create or develop and practice traditional knowledge.

Traditional Ecological Knowledge: Traditional knowledge relating to ecological systems acquired by communities living in close contact.

Traditional Knowledge: The knowledge, innovations and practices of indigenous people and local communities. It is the: knowledge gained through tradition from ancestors about plants and animals and agricultural practices and other matters.

Traditional Knowledge Innovations: Traditional Knowledge strictly refers to knowledge in public domain. TK innovations refer to knowledge that is not in public domain.

Traditional Medicine: It refers to health practices, approaches, knowledge and beliefs incorporating plant, animal and mineral based medicines, spiritual therapies, manual techniques and exercises, applied singularly or in combination to treat, diagnose and prevent illnesses or maintain well-being.

Traditional Resource Rights: A broad usage which include intellectual property rights, but denotes broader 'bundles of rights' including for example, human rights, land rights, religious rights, and cultural property.

B. Background Information

Indigenous Knowledge or Traditional Knowledge [TK] refers to the knowledge, innovations and practices of indigenous people and local communities. In other words, it is the: knowledge gained through tradition from ancestors about plants and animals, medicines, agricultural practices and other matters. Main points about traditional knowledge are listed below:

1. Traditional knowledge can be protected by *sui-generic* rights.
2. Traditional knowledge is also known as indigenous knowledge or local knowledge or traditional ecological knowledge.
3. It covers the matured long-standing traditions and practices of certain regional, indigenous, or local communities. In other words, it is gained from experience gained over the centuries and adapted to local culture and environment.
4. It includes knowledge about plants, animals, indigenous treatment of diseases through herbal medicines, cultural practices,*etc.*
5. Traditional knowledge is based on teachings and experiences and is orally passed on from generation to generation to some persons only. Earlier there were no written document on traditional knowledge.
6. Some forms of traditional knowledge are expressed through stories, legends, folklore, rituals, songs and even laws.
7. There is difference between local knowledge and traditional knowledge. The former existed from decade to century, whereas the latter existed from millennia.
8. It is collectively owned by the people or natives of an ecosystem or region.
9. Traditional knowledge incorporates knowledge of ecosystem and code of ethics governing appropriate use of environment. This code includes rules and conventions promoting desirable ecosystem relations, human-animal interactions and even social relationships. A combination of traditional and non-traditional knowledge will form a rich and distinctive understanding of life and the world.
10. The ecological knowledge is composed of both traditional knowledge and experiential knowledge [gained through personal experience].
11. Traditional knowledge is very important almost in every sphere of life. Hence, its protection is very necessary otherwise this valuable knowledge will be gradually lost.
12. The traditional knowledge has been classified into six major fields such as traditional knowledge of (i) agriculture, (ii) horticulture, (iii) forestry, (iv) medicines, (v) fisheries and human culture. These are briefly discussed as follows:

    (1) Agriculture: Traditional knowledge related to agriculture includes following aspects.

    (*i*) It includes knowledge of plant and animal genetic resources and their conservation.

(*ii*) Knowledge of production practices including development of plant varieties and animal breeds.

(2) Horticulture: Traditional knowledge related to horticulture includes following aspects.

(*i*) Knowledge of horticultural plant genetic resources and their conservation.

(*ii*) Knowledge of production and improvement of horticultural plant species.

(3) Forestry: It includes knowledge of timber yielding and other useful forest plants and their conservation.

(4) Medicines: It include knowledge of medicinal plant species which are used for treating human and animal diseases.

(5) Fisheries: It includes in depth knowledge about fish genetic resources and their conservation.

(6) Human Culture: It includes traditional knowledge which is expressed through stories, legends, folklore, rituals, songs and even laws.

Moreover, local people have better knowledge of environment such as snow, ice, weather, resources and the relationship between things. In other words, local people have better insight into the ecosystem of their region or locality.

13. **Registration:** Registration of traditional knowledge is not essential However, registration will provide legal rights to the people of local communities to protect their traditional knowledge.

14. **Duration:** The traditional knowledge can be protected for indefinite period by the local communities. In other words, there is no limit for conservation of traditional knowledge.

15. **Validity:** The traditional knowledge is general used in the area of its conservation by local people.

16. **Areas Covered:** The traditional knowledge is mainly of practical nature, particularly in such fields as agriculture, horticulture, forestry, fisheries, animals and herbal medicines. It also includes cultural matters such as stories, songs, folklore, proverbs, beliefs, cultural values, rituals, community laws and local language, *etc.*

17. **Transmission:** The knowledge is directly transmitted from individual to individual. Generally, it is passed on from father to son and mother to daughter generation after generation. Local people or natives possess both collective and individual traditional knowledge. The traditional knowledge is passed on by elders to the young. As the elders die, the some knowledge which has not been passed on is lost.

18. **Control:** The Native people have the rights to prevent use of their traditional knowledge by others without permission.

19. **Rights of T.K.:** The local people or natives, who have conserved the knowledge for centuries together, own the intellectual property rights to their traditional knowledge, even if much of it is yet to be written down. No one has the right to document or use traditional knowledge without permission. In other words, the following things are required for use of traditional knowledge:
    (*i*) Prior permission of natives
    (*ii*) The source of traditional knowledge should be acknowledged
    (*iii*) It should be properly interpreted.
    (*iv*) There should respect for traditional knowledge.

20. **Bio-piracy:** The illegal collection of indigenous plants by corporations who patent them for their own use is called bio-piracy. There is bio-piracy of traditional knowledge which needs to be prevented.

21. **Sharing of T.K.:** Local people possess both collective and individual traditional knowledge. Most traditional knowledge is shared among community members. But some traditional knowledge may be specific to an individual. This type of traditional knowledge is kept secret by some people who are engaged in treating some human and animal diseases. Moreover, it depends on the individual to decide whether he wishes to share his knowledge with outside parties or not.

22. **Approach:** Traditional knowledge assumes a holistic view including language, culture, practices, spirituality, mythology, customs and evn the social organization of local people. It can neither be divided nor separated from the local people who hold it. It is rooted in the spiritual health, culture and language of people. It is a way of life. In other words, it cannot be separated into secular[non-religious] and sacred knowledge.

23. **System:** Traditional knowledge is an authority system. It sets out the rules governing the use of resources, respect and an obligation to share. It is a dynamic, cumulative and stable system. Native people consider it truth.

24. **Adoption:** The traditional knowledge is adopted mainly in local areas where it has been developed. On the other hand, scientific knowledge is used in a larger area. In other words, traditional knowledge has limited practical applications as compared to scientific knowledge.

25. **Basis:** The explanation of traditional knowledge is based on examples, stories and myths, whereas the explanation of scientific knowledge is based on hypothesis, theories and law.

26. Many widely used products such as plant based medicines and cosmetics are derived from traditional knowledge.

27. Handicraft and non-wood forest products are based on TK.

28. Most indigenous and local communities live in areas which very rich in plant genetic resources. However, local people have made sustainable use of such resources for thousands of years. They have not exploited the resources.

29. The local people are rightly called as natural resource managers. They are managing plant and animal genetic resources for thousands of years. Women play a key role in the conservation of plant genetic resources as they are the primary harvesters of medicinal plants, seed stocks and small grains.

30. The indigenous people, as on site communities, have special skills and extensive knowledge of local environments. Their knowledge provides valuable information to the global community and useful model for biodiversity policies.

C. Multiple Choice Questions

1. TK comes under:

(a) Primary rights (b) Ordinary rights
(c) *Sui-generis* rights (d) None of these

2. TK includes knowledge of:

(a) Agriculture (b) Horticulture
(c) Fisheries (d) All of these

3. TK includes knowledge of:

(a) Medicines (b) Fisheries
(c) Human culture (d) All of these

4. TK includes knowledge of:

(a) Plants and animals (b) Herbal medicines
(c) Cultural practices (d) All of these

5. TK is based on:

(a) Examples (b) Stories
(c) Myths (d) All of these

6. On site managers of PGR refers to:

(a) Farmers (b) Breeders
(c) Scientists (d) All of these

7. **Scientific knowledge is based on:**
   (a) Hypothesis
   (b) Theories
   (c) Principles
   (d) All of these
8. **TK is based on:**
   (a) Teachings
   (b) Preaching
   (c) Experience
   (d) All of these
9. **TK is protected by:**
   (a) Local people
   (b) Scientists
   (c) Teachers
   (d) All of these
10. **Who has belief in TK?**
   (a) Doctors
   (b) Scientists
   (c) Local people
   (d) All of these

D. True or False

1. Indigenous property rights come under *sui-generis* rights.
2. Traditional knowledge is gained through personal experience.
3. Traditional knowledge is passed by elders to youngsters.
4. Generally traditional knowledge is passed from father to son.
5. Registration of traditional knowledge is not compulsory.
6. Local people are also known as indigenous people.
7. Indigenous people are onsite managers of plant genetic resources.
8. Community knowledge refers to indigenous knowledge.
9. Traditional knowledge cannot be separated from local people.
10. Traditional knowledge is orally passed on from one generation to the next.
11. Traditional knowledge is gained through traditions.

E. Fill in the Blanks

1. Indigenous property rights come under ________ rights.
2. Traditional knowledge is gained through ________ experience.
3. Traditional knowledge is passed on by ________ to youngsters.
4. Generally traditional knowledge is passed from father to ________.

5. Registration of traditional knowledge is not ________.

6. Local people are also known as ________ people.

7. Indigenous people are onsite ________ of plant genetic resources.

8. Community knowledge refers to ________ knowledge.

9. Traditional knowledge cannot be separated from ________ people.

10. Traditional knowledge is ________ passed on from one generation to the next.

11. Traditional knowledge is gained through ________.

F. Answer in Short or One Word

1. The unauthorized use of genetic resources and the associated knowledge by pharmaceutical companies.

2. Search for biological resources and related indigenous knowledge primarily for the purpose of commercial exploitation.

3. Knowledge held by communities, characterized by common or communal ownership.

4. Variety or multiple forms of human social structures, belief systems, and strategies for adapting to situations in different parts of the world.

5. Monumental remains of cultures such as humankind, the dramatic arts, languages and traditional music, and the informational, spiritual and philosophical systems upon which creations are based.

6. The practices, representations, expressions, as well as the knowledge and skills, that communities, groups and, in some cases, individuals recognize as part of their cultural heritage.

7. The knowledge held by communities and peoples that are indigenous.

8. People whose ancestors inhabited a place or country.

9. A core proposition is that future generations have a right to an inheritance (capital bequest) sufficient to allow them to generate a level of well-being no less than that of the current generation.

10. A non-profit making, voluntary, service-oriented/development oriented organization.

11. Consent that is required for accessing genetic resources or the associated knowledge held by communities living in the locality.

12. The knowledge gained through education and scientific readings.

13. Development that meets the needs of the present without compromising the ability of future generations to meet their own needs.

14. Persons who create or develop and practice traditional knowledge.

15. Traditional knowledge relating to ecological systems acquired by communities living in close contact.

16. The knowledge, innovations and practices of indigenous people and local communities.

17. Knowledge that is not in public domain.

18. Health practices, approaches, knowledge and beliefs incorporating plant, animal and mineral based medicines, spiritual therapies, manual techniques and exercises, applied singularly or in combination to treat, diagnose and prevent illnesses or maintain well-being.

19. Rights related to human rights, land rights, religious rights, and cultural property.

ANSWERS

C. Multiple Choice Questions

Sl.No.	*Answer*	*Sl.No.*	*Answer*
1.	(c) *Sui-generis* rights	**6.**	(a) Farmers
2.	(d) All of these	**7.**	(d) All of these
3.	(d) All of these	**8.**	(d) All of these
4.	(d) All of these	**9.**	(a) Local people
5.	(d) All of these	**10.**	(c) Local people

D. True or False

1 to 11 all are true

E. Fill in the Blanks

Sl.No.	*Answer*	*Sl.No.*	*Answer*
1.	*Sui-generis*	**7.**	Managers
2.	Personal	**8.**	Indigenous
3.	Elder	**9.**	Local
4.	Son	**10.**	Orally
5.	Compulsory	**11.**	Traditions
6.	Indigenous		

F. Short or One Word Answer

Sl.No.	*Answer*	*Sl.No.*	*Answer*
1.	Bio-piracy	**11.**	Prior informed consent
2.	Bio-prospecting	**12.**	Scientific knowledge
3.	Community knowledge	**13.**	Sustainable development
4.	Cultural diversity	**14.**	Tradi. knowledge holder
5.	Cultural heritage	**15.**	Tradi. ecological knowledge
6.	Intangible heritage	**16.**	Traditional knowledge
7.	Indigenous knowledge	**17.**	Traditional knowledge innovation
8.	Indigenous people	**18.**	Traditional medicines
9.	Intergenerational equity	**19.**	Traditional resource rights
10.	NGO		

Chapter 21

Matching Type Questions

Q. 1: Match the information of Group A with Group B:

Sl.No.	*Group A*	*Sl.No.*	*Group B*
1.	Land	**a.**	Movable property
2.	Gold	**b.**	Immovable property
3.	Invention	**c.**	Status
4.	Property	**d.**	Intellectual property

Answer: 1-b, 2-a, 3-b, 4-c.

Q. 2: Match the information of Group A with Group B:

Sl.No.	*Group A*	*Sl.No.*	*Group B*
1.	House	**a.**	Movable property
2.	Animals	**b.**	Immovable property
3.	Copyright	**c.**	Invention
4.	Patent	**d.**	Books

Answer: 1-b, 2-a, 3-d, 4-c.

Q. 3: Match the information of Group A with Group B:

Sl.No.	*Group A*	*Sl.No.*	*Group B [Protection in years]*
1.	Full term patent	**a.**	6-10 years
2.	Short term patent	**b.**	20 years
3.	Utility Patent	**c.**	5 years
4.	Industrial design	**d.**	10 years

Answer: 1-b, 2-d, 3-a, 4-c.

Q. 4: Match the information of Group A with Group B:

Sl.No.	*Group A*	*Sl.No.*	*Group B [Protection in years]*
1.	Copyrights	**a.**	05
2.	Patent	**b.**	10
3.	Trademarks	**c.**	20
4.	Industrial design	**d.**	60

Answer: 1-d, 2-c, 3-b, 4-a.

Q. 5: Match the information of Group A with Group B:

Sl.No.	*Group A*	*Sl.No.*	*Group B [Protection in years]*
1.	Utility models	**a.**	10
2.	Geographical indications	**b.**	6-10
3.	Trade secret	**c.**	15
4.	Performers	**d.**	Indefinite

Answer: 1-b, 2-a, 3-d, 4-c.

Q. 6: Match the information of Group A with Group B:

Sl.No.	*Group A*	*Sl.No.*	*Group B [Protection in years]*
1.	Performers	**a.**	20
2.	Film producers	**b.**	15
3.	Broadcasting organization	**c.**	6-10
4.	Utility models	**d.**	50

Answer: 1-b, 2-d, 3-a, 4-c.

Q. 7: Match the information of Group A with Group B:

Sl.No.	*Group A*	*Sl.No.*	*Group B [Protection in years]*
1.	Photographers	**a.**	20
2.	Full term patent	**b.**	25
3.	Semiconductor designers	**c.**	01
4.	Domain name	**d.**	10-15

Answer: 1-b, 2-a, 3-d, 4-c.

Q. 8: Match the information of Group A with Group B:

Sl.No.	*Group A*	*Sl.No.*	*Group B [Protection in years]*
1.	Trade name	**a.**	10
2.	Registered Mask	**b.**	15

3.	Unregistered Mask	**c.**	03
4.	Database rights	**d.**	02

Answer: 1-3, 2-c, 3-d, 4-b.

Q. 9: Match the information of Group A with Group B:

Sl.No.	*Group A*	*Sl.No.*	*Group B [Protection in years]*
1.	PBR perennial crops	**a.**	Till author is alive
2.	Moral rights	**b.**	18 years
3.	Supplementary protection certificate	**c.**	Indefinite
4.	Traditional knowledge	**d.**	05 years

Answer: 1-b, 2-a, 3-d, 4-c.

Q. 10: Match the information of Group A with Group B:

Sl.No.	*Group A*	*Sl.No.*	*Group B*
1.	Bansmati Rice	**a.**	Natural GI
2.	Darjeeling tea	**b.**	Manufactured GI
3.	Agra petha	**c.**	Agricultural GI
4.	Copyrights	**d.**	Book

Answer:1-c, 2-a, 3-b, 4-d.

Q. 11: Match the domain extension with their use

Sl.No.	*Domain Extension*	*Sl.No.*	*Used fo*
1.	.com	**a.**	Aviation
2.	.net	**b.**	Commercial organizations
3.	.org	**c.**	Internet service
4.	.aero	**d.**	Non-profit organizations

Answer:1-b, 2-c, 3-d, 4-a.

Q. 12: Match the domain extension with their use

Sl.No.	*Domain Extension*	*Sl.No.*	*Used fo*
1.	.biz	**a.**	Cooperatives
2.	.coop	**b.**	Business
3.	.pro	**c.**	Museum
4.	.museum	**d.**	Professionals

Answer:1-b, 2-a, 3-d, 4-c.

Q. 13: Match the information of Group A with that of Group B

Sl.No.	*Particulars*	*Sl.No.*	*Period of Protection*
1.	Trees and vines: Minimum period	**a.**	6 years
2.	Trees and vines: Maximum period	**b.**	15 years
3.	Field crops: Minimum period	**c.**	9 years
4.	Field crops: Maximum period	**d.**	18 years

Answer:1-c, 2-d, 3-a, 4-b.

Q. 14: Match the information of Group A with that of Group B

Sl.No.	*Particulars*	*Sl.No.*	*Period of Protection*
1.	First UPOV Act	**a.**	2001
2.	Second UPOV Act	**b.**	1989
3.	Farmers rights' Act	**c.**	1978
4.	PVP and FR Act	**d.**	1991

Answer:1-c, 2-d, 3-b, 4-a.

Appendices

Appendix 1: Summary of Intellectual Property Rights

Sl.No.	*Type of IPR*	*Registration*	*Period Years*	*Authori-zation*	*Requirements*
A.	**Primary Rights**				
1.	Copyrights	Not essential	70	Yes	Originality Fixed expression
2.	Patents	Essential	20	Yes	Novelty Usefulness Non-obviousness
3.	Trademarks	Not essential	10	Yes	Distinctiveness No multiple appearances
4.	Industrial Design	Essential	05	Yes	Novelty Commercial use
5.	Utility Models	Essential	6-10	Yes	Novelty Industrial application
6.	Geographical Indications	Not essential	10	No	Special quality Reputation
7.	Trade Secret	Not essential	Indefinite	Yes	Confidential Commercial use
8.	Related Rights Performers and Database creators	Not essential	15	Yes	Originality Usefulness
	Phonogram producers and Broadcasting	Not essential	20	Yes	Originality Usefulness

	Photographers and designers	Not essential	25	Yes	Originality Usefulness
	Film Producers	Not essential	50	Yes	Originality Usefulness
9.	Domain Names	Essential	01	Yes	Unique Internet Web address
10.	Trade Names	Not essential	03	Yes	Novelty Distinctiveness
B.	***Sui-generis* Rights**				
11.	Database Rights	Not essential	15	Yes	New creation and Investment
12.	Mask Work Rights Registered Unregistered	Not essential	 10 02	Yes	Originality and Usefulness
Sl.No.	*Type of IPR*	*Registration*	*Period Years*	*Authori-zation*	*Requirements*
13.	Plant Breeders' Rights Annual Plants Perennial Plants	Essential	 15 18	Yes	Novelty Distinctiveness Uniformity Stability
14.	Moral Rights	Not essential	Lifelong	Yes	Linked with copyrights
15.	SPC	Essential	05	Yes	After expiry of basic patent
16.	Traditional Knowledge	Not essential	Indefinite	No	Large scale usefulness

Appendix 2: Address of Patent, Copyright and Trademark Offices in India.

Sl.No.	*Name of IPR*	*Head Office/Branch Office*	*Location*
1.	Patent	Head Office	Nizam Palace, Kolkata.
		Branch Unit	Karol Baugh, New Delhi.
		Branch Office	Parel West, Mumbai.
		Branch Office	Wallajah Road, Chennai.
		Patent Information System	CGO Complex, Seminary Hills, Nagpur.
2.	Patents, Designs and Trademarks	Controller General of Patents, Designs and Trademarks	Maharishi Kurvey Road, Mumbai.
3.	Copyrights	Registrar of Copyrights	Ministry of HRD, Kasturba Gandhi Marg, New Delhi.
4.	Trademarks	Head Office	Old Central Building, M.K. Road, Mumbai.
		Branch Office	Ashram Road, Ahmedabad.
		Branch Office	Chowringhee Square, Kolkata.
		Branch Office	Okhla Industrial Estate, New Delhi.
		Branch Office	Basant Nagar, Chennai.

Appendix 3: Country Code Top Level Domain [ccTLD] Extensions for some selected Countries.

Sl.No.	*Country*	*Domain Name*	*Sl.No.*	*Country*	*Domain Name*
1.	Austria	.at	26.	Myanmar	.mm
2.	Australia	.au	27.	Nepal	.np
3.	Argentina	.ar	28.	Netherlands	.nl
4.	Bangladesh	.bd	29.	New Zealand	.nz
5.	Belgium	.be	30.	Nigeria	.ng
6.	Brazil	.br	31.	Norway	.no
7.	Cambodia	.kh	32.	Pakistan	.pk
8.	Canada	.ca	33.	Peru	.pe
9.	China	.cn	34.	Philippines	.ph
10.	Colombia	.co	35.	Poland	.pl
11.	Cuba	.cu	36.	Saudi Arabia	.sa
12.	Denmark	.dk	37.	South Africa	.za
13.	Egypt	.eg	38.	Spain	.es
14.	Finland	.fi	39.	Sri Lanka	.lk
15.	France	.fr	40.	Singapore	.sg
16.	Germany	.de	41.	Sweden	.sd
17.	Greece	.gr	42.	Switzerland	.ch
18.	Hungry	.hu	43.	Thailand	.th
19.	India	.in	44.	Turkey	.tr
20.	Indonesia	.id	45.	Uganda	.ug
21.	Israel	.il	46.	UAE	.ue
22.	Italy	.it	47.	United Kingdom	.uk
23.	Japan	.jp	48.	USA	.us
24.	Malaysia	.my	49.	Vietnam	.vn
25.	Mexico	.mx	50.	Zambia	.zm

Glossary

A

Abandonment: An application for trademark or copyright registration may be considered abandoned if the applicant does not take steps to complete the process. An application for Industrial design will be considered abandoned if the applicant fails to reply within the specified time to any report that sets out objections to registration.

Abstract: A brief summary of an invention.

Abuse of Patent: Abusing one's patent rights by failing to make one's invention commercially available.

Advertisement: Publication of a trademark application in the *Trademarks Journal*. Details are published to allow opportunity for challenges to the application (opposition).

Allowance: Acknowledgement by the Trademarks Office that an application is eligible for registration.

Application: The formal request for registration of intellectual property.

Artistic Work: Any visual representation such as a painting, drawing, map, photograph, sculpture, engraving or architectural plan.

Assignment: Transfer of intellectual property rights from the owner to another party.

Author: The creator of an artistic, literary, musical or dramatic work.

B

Berne Copyright Convention: An international treaty extending copyright protection in member countries to nationals of other member countries. Canada is a signatory to this treaty.

Biodiversity: Variety of biological organisms found in the entire world or it the some of genes, species and ecosystem in a region.

Breeders' Exemption: Rights provided to Plant Breeders use protected material as the basis to develop a new variety or for other research work; also called research exemptions or breeders' privilege.

C

Cancopy: A reprography collective that grants licenses to photocopy works and collects fees on behalf of its members.

CBD: Convention on Biological Diversity

Certificate of Registration: Official confirmation that your design, copyright or trademark has been registered.

Certification marks: Marks identifying goods or services meeting a defined standard (*e.g.*, Wool Mark on clothing).

Claims: The basis for filing an application that allows for the determination of the person entitled to the registration of the mark.

Clearly Descriptive: A mark that clearly describes a feature of a ware or service and that is therefore not registered as a trademark.

Coexistence Agreement: A binding agreement whereby two or more traders agree how to operate in the same or similar market to avoid any potential confusion.

Commercial Cultivar: A variety which is used for cultivation on vast area.

Candidate Variety: A variety which is to be ptotected under Plant Variety Protection Act.

Copyright: A state monopoly granted to protect artistic, musical and literary works.

Collective: Organization that administers rights granted by the copyright system on behalf of copyright owners who have joined that collective.

Compulsory License: The right granted by the Commissioner of Patents to one or more parties to produce a patented invention. These are granted under certain circumstances as a measure against abuse of patent rights.

Copyrights: Rights that Provide protection for literary, artistic, dramatic or musical work, computer programs, and sound recordings.

Copyright Act: The legislation governing copyright in in a country.

Copyright Declaration: Formal statement that the applicant is the author, the owner of the work, an assignee or a person to whom an interest in the copyright has been granted by license.

Copyright Infringement: Violation of copyright through unauthorized copying or use of a work or other subject matter under copyright.

Copyright Office: The central government office responsible for registering copyrights. **Crown copyright:** Copyright of works prepared for or published by the government, *i.e.*, government publications.

D

Database: A collection of data organized for rapid search and retrieval by a computer

Deceptively Mis-descriptive: A mark that deceptively mis-describes a feature of a ware or service in such a way as to potentially mislead the consumer is not registered as a trademark.

Declaration: Formal statement that you are the proprietor of the design and that, to your knowledge, no one else was using the design when you created it.

Description: A description is a basic requirement of an application for an industrial design. It identifies the features of shape, configuration, pattern or ornament that distinguish the design, and indicates whether the design resides in the entire article or a portion.

Design: The appearance of the whole or part of a product, including any pattern or texture applied to a surface.

Design Patent: The term used in the United States to describe a form of protection similar to that of industrial design protection in Canada.

Disclaimer: A statement that the applicant does not claim exclusive rights for a portion of the trademark.

Distinctiveness: The variety must be distinguishable in at least one character from previously available varieties.

Distinguishing: The unique shape of a product or package that can be registered as a trademark.

Domain Name: The virtual address on the Internet.

Dramatic Work: Includes plays, screenplays, scripts, films, videos and choreographic works as well as translations of such works.

Due Diligence: Investigations undertaken to assess the strength of one or more intellectual property rights.

Drawings: Drawings (or photographs) disclose the industrial design and are a basic requirement of a design application.

E

Ecosystem Diversity: Biological variation found in the ecosystems in a country.

EPO (European Patent Office): The pan-European body responsible for granting patents across Europe.

Examination

(i) **Trademarks:** The process through which the Trademarks Office determines whether an application for trademark is registrable.

(ii) **Industrial Designs:** The process through which the Industrial Design Office determines whether a design warrants registration. The main objective is to determine whether the design is indeed design subject matter, that it is original, and that the documentation meets the requirements of the *Acts* and *Rules*.

Exception: A provision in a copyright law permitting the use of a work by defined user groups without the consent of its creator or without the payment of royalties, conditions that would normally constitute an infringement of copyright. Examples of user group beneficiaries from exceptions are educational institutions, libraries, museums, archives and persons with a perceptual disability.

Example Variety: In DUS, testing, a variety which is used for comparing a particular trait.

Exotic Variety: A foreign variety which is directly recommended for commercial cultivation.

Extant Variety: All released and notified varieties which have not been protected under plant variety protection act. It includes notified varieties, private varieties, farmers' varieties and varieties of common knowledge.

F

Fair Dealing: Use of works for purposes of private study, research, criticism, review or news reporting that does not constitute infringement of copyright.

Farmers' Exemption: Legal rights provided to farmers to save, use, exchange, share or sell his farm produce of a protected variety; also called farmers' rights or farmers' privilege.

Farmers' Rights: The legal rights provided to farmers to save, use, sow, replant, exchange, share or sell his farm produce including seed of a variety protected under Plant Variety Protection Act

Farmers Variety: A variety developed by farmers and used for cultivation for several years.

Fees: Specific money payable for various services provided by Intellectual Property Office.

First to File: A patent system in which the first inventor to file a patent application for a specific invention is entitled to the patent. In most of the countries, the first person to file has priority over other people claiming rights for the same invention.

G

Genetic Diversity: The variety of genes and genotypes found in a particular species.

Globalization: The process of organizing the whole world into a single integrated marketing unit.

Green Revolution: Quantum jump in the productivity of cereals by evolving high yielding varieties.

I

Improved Cultivars: Varieties developed by scientific method and released for commercial cultivation by central or State variety release committee.

Industrial Design: The visual features of shape, configuration, pattern or ornament applied to a manufactured article.

Industrial Design Act: Legislation governing registration and ownership of industrial designs in a country.

Industrial Design Office: The central government office responsible for registration of industrial designs.

Infringement: The unauthorized copying or use of a protected form of intellectual property. It may include unauthorized use of industrial design, trademarks and patents.

Industrial Design: Violation of industrial design rights through unauthorized use of a design.

- **(i) Trademarks:** Violation of trademark rights through unauthorized use of a trademark.
- **(ii) Patents:** Unauthorized use or sale of a protected invention.

Integrated Circuit Topography: (a) The three-dimensional configuration of the electronic circuits embodied in integrated circuit products or layout designs. (b) Legal protection against imitation of this configuration.

Intellectual Property: A form of creative work that can be protected through a trademark, patent, copyright, industrial design or integrated circuit topography.

Intellectual Property Office: The government agency responsible for the administration of intellectual property laws.

Intellectual Property Right: Legal rights granted to protect a form of intellectual property for *e.g.* a patent right, design right, trademark right or copyright.

Intellectual Property Assets: Those intangible assets that derive from intellectual property and that, typically, add value to your business.

Intellectual Property Auditing: Investigations that are undertaken to determine the nature and number of Intellectual Property Rights owned by a business.

Invention: An innovation that is both novel and non-obvious.

L

Landraces: The primitive cultivars which were selected and cultivated by farmers for many generations.

Letter of Consent: Letter from one party, typically the owner of a stronger trademark, consenting to the use and/or registration of another party's trademark.

Literary Work: Work consisting of text such as novels, poems, song lyrics without music, catalogue, reports, tables as well as translations of such works. It also includes computer programs.

License: Legal agreement granting permission to any person to use Intellectual Property Rights in exchange for some consideration or benefit. A license does not change ownership of the copyright.

License: A person other than the owner having the same rights as the owner for use and sale of a product, mark or design.

Licensing Design: Granting of an exclusive license from the owner of the design to another party that allows said party to manufacture, use and sell the design, and to sub-license others to do so.

Licensing Invention: Allowing a business or individual to manufacture and sell an invention, usually in exchange for royalties.

M

Madrid Protocol: An international agreement for the registration of trademarks in a number of countries by means of a single application.

Mask **Work: A** two or three-dimensional layout or topography of an integrated circuit (IC or "chip").

Maintenance Fee:

- **(i) Industrial Design:** Fee required in order to maintain the rights to an for a second five year period.
- **(ii) Patents:** Annual fee to keep a patent.

Marking:

- **(i) Copyrights:** Indicating copyright with a small "c" in a circle (©), the name of the copyright owner and the year of first publication.
- **(ii) Industrial design:** For registered design a capital "D" is placed in a circle along with the name, or abbreviation thereof, of the design's proprietor on the object to which the design has been applied, or to its label or packaging.
- **(iii) Trademarks:** Trademark owners often indicate their registration through certain symbols, such as R in a circle (registered), TM (trademark), SM (service mark).

Modern Cultivars: The currently cultivated high yielding varieties.

Monopoly Right: Another term for Intellectual Property Right; the period during which a patent, design, trademark or copyright can be exclusively exploited by its owner.

Moral Rights: Rights an author retains over the integrity of a work and the right to be named as its author even after sale or transfer of the copyright.

Musical Work: Work that consists of music plus lyrics or music only.

N

Nom-de-plume: A name used by a writer instead of his or her real name.

Novelty: The newness of a invention, product, marks, variety, *etc.*

O

OHIM (Office for Harmonization in the Internal Market): The European Community's Trade Mark Registration Authority.

Opposition: The process whereby members of the public may object to the granting of a trademark registration if they have valid grounds for doing so.

P

Patent: A legal right provided to the inventor to exclude others from making, using or selling his invention.

Patent Agent: A fully qualified under the rules of the UK's Chartered Institute of Patent Agents.

Patent Application: A technical document describing in detail an innovation for which a patent is sought.

Patent Attorney: A fully qualified to act before the Patent Office in a country.

Patent Cooperation Treaty: An international treaty providing for standardized filing procedures for foreign patents in the countries that have signed the treaty.

Patent Examiner: An official with technical expertise charged with the task of classifying a patent or an application, or of determining whether a patent application fulfills the requirements for a grant of patent.

Patent Office: The patent granting authority and disseminator of patent information.

Patent Office Action: An official written communication by the Patent Office on the merits of an application.

Patent Pending: A label sometimes affixed to new products informing others that the inventor has applied for a patent and that legal protection from infringement (including retroactive rights) may be forthcoming.

Performers and Producers Rights: A term used to indicate rights of performers and sound recording producers to be remunerated when their performances and sound recordings are performed publicly or broadcast. Also referred to as "neighbouring rights".

Performing Rights Society: An organization that administers rights to perform musical works on behalf of composers, lyricists, songwriters and music publishers.

Piracy: An unauthorized duplication of goods protected by intellectual property law (*e.g.* copying software unlawfully).

Plagiarism: Using the work (or part of it) of another person and claiming it as your own.

Posthumous Work: A work that is published for the first time (or, for certain types of works, published, performed or delivered in public for the first time) after the author's death.

Popular Variety: A widely grown crop variety.

Preferred Embodiment: An explanation, in a patent application, of the most appropriate and useful practical applications of an invention.

Primitive Cultivars: Same as land races.

Prior Art (Industrial Design): The body of registered industrial designs or other published designs including catalogues, manuals and other publications throughout the world relevant to the application for registration of a design.

Private Copying: Copying of pre-recorded musical works, performers' performances and sound recordings into a blank medium, such as audio tape or cassette, for personal use.

Private Copying Levy: An amount of money established by the Copyright Board on blank recording media, whether made here or imported, by manufacturers and importers and is then sold in Canada redistributed to eligible composers, lyricists, performers and sound recording makers through their professional associations and collectives.

Prosecution: All the steps involved in completing a patent application.

Proprietor: The proprietor of an industrial design is the owner of the design. This can be the person who created it or who paid to have it created or who bought the design rights from the original owner.

Publication: Making available copies of a work to the public.

Plant Breeders' Rights: Legal rights granted to the Plant Breeders/owner for the control over the multiplication and sale of their particular plant variety.

Place of Origin: A word or depiction that designates the origin of a product or service and that therefore may not be registered as a trademark.

Preliminary Search: The search of Trademarks Office records that one should undertake before submitting an application for trademark registration.

Pre-publication Verification: A second search of Trademarks Office records to ensure that no confusing trademarks exist before a given application is published in the *Trademarks Journal*.

Prohibited Marks: Marks that are specifically prohibited from use in the Trademarks Act.

Proposed Use: A statement indicating intention to use a certain trademark and how that use will occur.

R

Registration: The granting of formal recognition of a trademark, copyright, industrial design, or integrated circuit topography by the Trademarks, Copyright, and Industrial Design Office respectively.

Registered Patent Agent: A specialist entitled to prepare and prosecute patent applications.

Registered Trademark: A trademark entered on the federal government's Trademarks Register, which formally recognizes the owner's rights to the mark.

Registrar of Trademarks: The official responsible for the Trademarks Office in a country.

Registrar of Copyrights: The official responsible for the Copyright Office in a country.

Registrar of Integrated Circuit Topographies: The official responsible for the Integrated Circuit Topography Office in a country.

Renewal Fee:

(i) **Industrial Designs:** Fee required in order to renew the rights to an industrial design for a second five-year period.

(ii) **Trademarks:** Fee required in order to renew a trademark registration for an additional fifteen-year period.

Representative for Service: A person appointed by the applicant or registered proprietor to receive documents on his/her behalf.

Rome Convention Country: A country that is party to the international convention for the protection of performers, producers of phonograms and broadcasting organizations done at Rome on October 26, 1961.

Royalty: A sum paid to copyright owners for the sale or use of their works, or other subject matter.

S

Search: The act of searching through intellectual property records in order to verify whether a patent, trademark or industrial design has been previously filed or registered.

SOCAN: The Society of Composers, Authors and Music Publishers of Canada. SOCAN is the only performing rights society in Canada.

Sound Recording: A device which produces sounds, such as cassette, record, or CD player.

Special Order Examination: A request to the Patent Office for early consideration of a patent application.

Specification: Part of the patent application. It includes a detailed description of the invention, claims specifying the aspects of the invention for which protection is sought, and the extent of the protection being sought

Stability: Each generation of the variety should be similar.

Supplementary Protection Certificate (SPC): A special form of intellectual property (IP) that extends the protection of a patented active ingredient or combination of active ingredients present in a pharmaceutical or plant protection product after the expiry of the patent.

T

Tariff: A standard charge for use of copyrighted works. Usually applies to fees paid by users of musical works and cable companies for the rebroadcast of programs.

Trademark: A word, symbol or design, or combination of these, used to distinguish the goods or services of one party from those of another.

Trade Secret: Information about a product or process kept secret from competitors.

Trademark Agent: A a person whose name is entered on the list of trademark agents and who is therefore entitled to practice before the Trademarks Office.

Trademarks Act: Central government legislation governing trademark registration in a country.

Trademarks Journal: A publication of the Trademarks Office issued weekly and containing all approved applications and Office rulings.

Trademarks Office: The central government agency responsible for registering trademarks in a country

Trademarks Records: The inventory of registered trademarks and pending applications maintained by the Trademarks Office.

Trademarks Register: The official listing of registered trademarks.

Trade Name: The name under which a company conducts its business. Trade names may or may not be considered trademarks, depending on the circumstances.

Traditional Knowledge [TK]: The knowledge, innovations and practices of indigenous people and local communities

U

Uniformity: All plants in a variety should look alike.

Universal Copyright Convention: A country that is party to the Universal Copyright Convention [adopted on September 6, 1952 in Geneva, Switzerland, or to that convention as revised in Paris, France, on July 24, 1971].

V

Variety: A genotype released for commercial cultivation either by State Variety Release Committee or by Central Variety Release Committee and notified by government.

W

WIPO (World Intellectual Property Organization): International body responsible for facilitating the granting of Intellectual Property Rights through out the world.

WTO Member Country: A Country that is a member of the World Trade Organization as defined in subsection 2(1) of the *World Trade Organization Agreement* Implementation Act.

World Trade Organization [WTO]: An international organization that regulates trade and tariffs between nations in order to ensure that trade flows smoothly, predictable and as freely as possible.

References

Alexander I., Poltorak and Paul, J. L. 2002. Essentials of Intellectual Property. John Wiley and Sons.

Alexander I., Poltorak and Paul, J. L. 2004. Essentials of Licensing Intellectual Property. John Wiley and Sons.

Amrita Narlikar 2005. The World Trade Organization. A Very Short Introduction. Oxford Uni. Press, USA.

Andrew, H. J., Phill, G. and Terri, R. G. 2006. Plant Diversity.

Arthur, R. M. and Michael, H. D. 2000. Intellectual Property: Patents, Trademarks and Copyrights [Nutshell Series]. West Group Publishing.

Bhagwati, J. 2005. In Defense of Globalization. Oxford Uni. Press, USA.

Carlotte Fiell and Peter Fiell 2000. Industrial Design A-Z. Taschen.

Carrie Russell 2004. Complete Copyright. An Everyday Guide for Librarians. American Library Association.

Clarisa, L. (ed) 2000. Intellectual Property Rights in Emerging Markets. Aei Pr.

David, J. Moser 2001. Music Copyright for the New Millenium. Artist pro. 1st edn.

David, Pressman. 2005. Patent It Yourself. 11th edn. NOLO.

Donald, A. Norman 2002. The Design of Everyday Things. Basic Books.

Gary, P. Sampson 2001. The Role of the World Trade Organization in Global Governance. UN Uni. Press.

Gert, W. R., Paul, V. K. and Bart, K. M. E. 2006. European Community Plant Variety Protection. Oxford Uni. Press.

Glenn Adamson 2003. Industrial Strength Design: How Brooks Stevens Shaped Your World ?. The MIT Press.

Henri Charmasson 2004. Patents, Copyrights and Trademarks for Dummies. For Dummies.

Henry, R. J. 2004. Plant Diversity and Evolution: Genotypic and Phenotypic Variation in Higher Plants. Southern Cross Uni., Australia.

Jessica Litman 2006. Digital Copyrights. Prometheus Books.

Jill Gilbert 2004. The Entrepreneur's Guide to Patents, Copyrights, Trademarks, Trade secrets and Licensing. Berkley Trade.

Joanna, R.J. 1997. Merchandising Intellectual Property Rights. John Wiley and Sons.

John Heskett 1985. Industrial Design: World of Art. Thames and Hudson.

Joseph, E. S 2006. Making Globalization Work. W. W. Norton.

Laurence R. Helfer 2005. Intellectual Property Rights in Plant Varieties: International Legal Regimes and Policy Options. FAO-UN.

Lee Wilson 1998. Trademark Guide. Allworth Press.

Lynn, H. and Cheryl, D. C. 2004. Design Secrets: Products 2: 50 Real Life Projects Uncovered. Rockport Publishers.

Mitsuo, M., Thomas, J. S. and Patros, C. M. 2006. The World Trade Organization: Law, Practice and Policy. Oxford Uni. Press, USA.

Patterson, L. R. and Stanley, W. L. 1991. The Nature of Copyright: A Law of Users Rights. Universal Georgia Press.

Paul Gouldstein 2004. Copyrights, Patents, Trademarks and related State Doctorines Foundation Press.

Peter Naray 2001. Russia and the World Trade Organization. Palgrave Macmillen.

Raymond Loewy 2000. Industrial Design. Overlook TP.

Richards, S. 2006. Patent, copyright and Trademark. An Intellectual Property Desk Reference. Amazon Book.

Richard, W. Stim 1999. Copyright Law. Thompson Delmar Learning 1st Edn.

Robert, A. G. and Jane, C. G. 2002. Copyright: Cases and Materials. Foundation Press, New York.

Robert C. Lind 2000. Copyright Law. Carolina Academic Press.

Robert, W. H. 2005. Intellectual Property Rights in Frontier Industries. Aei Pr.

Singh, Phundan 2010. IPR and Plant Breeders' Rights. New Vishal Publications, New Delhi.

Siva Vaidhyanathan 2003. Copyrights and Copy wrongs: The Rise of Intellectual Property and How it Threatens Creativity. New York Uni. Press.

Stephen Elias 2003. Trademark: Legal Care for your Business and Product Name. NOLO 6th edn.

Stephen Fishman 2001. Copyright Your Software. 3rd edn, NOLO.

Stephen Fishman 2006. Copyright Hand Book. What Every Writer Needs to Know? NOLO.

Books by the Same Author

A. Plant Breeding

1. Essentials of Plant Breeding
2. Plant Breeding: Molecular and New Approaches
3. Molecular Plant Breeding
4. Principles of Seed Technology
5. Seed Technology: At a Glance
6. Objective Seed Technology
7. Practical in Crop Breeding
8. Practical and Numerical in Plant Breeding
9. Numerical Problems in Plant Breeding and Genetics
10. Objective Science of Plant Breeding
11. Plant Breeding At a Glance
12. Plant Breeding [For Under Graduate Students]
13. Principles of Plant Breeding
14. Cotton Breeding
15. Heterosis Breeding in Cotton
16. Breeding Hybrid Cotton
17. Breeding Transgenic Bt. Cotton
18. Cotton Improvement in India
19. Glossary of Plant Breeding and Genetics
20. IPR and Plant Breeders' Rights [Subjective]
21. IPR and Plant Breeders' Rights [Objective]
22. IPR and Plant Breeders' Rights [At a Glance]
23. Introduction to Intellectual Property Rights

24. Intellectual Property Rights at a Glance
25. Intellectual Property Rights: Objective Analysis
26. Elements of Baby Corn
27. Breeding Crop Plants for Stress Resistance
28. Plant Breeding: Related Legislations

B. Genetics

29. Elements of Genetics
30. Genetics
31. Principles of Genetics [For UG Students]
32. Plant Genetics
33. Cotton Genetics
34. Objective Genetics
35. Genetics At a Glance
36. Objective Genetics and Plant Breeding
37. Objective Genetics and Plant Breeding [Hindi Edn]
38. Glossary cum Dictionary of Genetics and Plant Breeding
39. Molecular Genetics [Objective]
40. Molecular Genetics [At a Glance]
41. Molecular Genetics [Subjective]
42. Genetics and Man
43. Genetics Today

C. Quantitative Genetics

44. Biometrical Techniques in Plant Breeding
45. Application of Biometrical Techniques in Plant Breeding [Hindi Edn]
46. Objective Quantitative Genetics
47. Quantitative Genetics At a Glance
48. Quantitative Genetics

D. Biotechnology

49. Introduction to Biotechnology
50. Plant Biotechnology
51. Principles of Plant Biotechnology
52. Plant Biotechnology: At a Glance
53. Objective Plant Biotechnology

E. Competitive Examination, *etc.*

54. A Guide to Competitive Examinations of Agriculture
55. Objective General Agriculture
54. Glimpses of Cotton Breeding

Index

L

M

P

R

S

T

U